MORE SELF THAN SELF

MORE SELF THAN SELF
At Autism's Edge

HENRY KONG M.D.

iUniverse, Inc.
New York Lincoln Shanghai

More Self than Self
At Autism's Edge

iUniverse books may be ordered through booksellers or by contacting:

iUniverse
2021 Pine Lake Road, Suite 100
Lincoln, NE 68512
www.iuniverse.com
1-800-Authors (1-800-288-4677)

ISBN-13: 978-0-595-39296-4 (pbk)
ISBN-13: 978-0-595-83689-5 (ebk)
ISBN-10: 0-595-39296-2 (pbk)
ISBN-10: 0-595-83689-5 (ebk)

Printed in the United States of America

In memory of Hans Asperger (1906-1980)
and to the millions around the world who share his syndrome.

For Sir Francis Crick (1916-2004), the greatest biologist of the twentieth century. He made DNA plausible.

For Professor Jessica Treisman, the greatest biologist of the twenty-first century. She made this book possible.

For my parents. They made me probable.

Contents

Foreword

For those who have wondered what it feels like to get under the skin of someone with autism or Asperger's syndrome, this first book by Dr. Henry Kong will convey the experience. Personal vignettes are interspersed with explanations of the science underlying the various defects associated with Asperger's syndrome. His fascinating account of how the brain accomplishes perception, motor coordination, emotion and decision making sheds light on normal function as well as suggesting what might have gone wrong in the autistic brain. He effectively conveys the excitement of the latest discoveries in neuroscience to the lay reader. Dr. Kong takes his analysis to the genetic as well as the anatomical and behavioral levels, and even discusses how evolution has shaped our minds and brains. In one chapter he mentions the savant skills and obsessive mental pursuits of some autistics; it seems that his own obsession has been effectively directed towards understanding the cause of his disorder.

Dr. Kong's honest description of the difficulties he faced growing up not only as a high-functioning autistic, but also as an Asian immigrant to the US with an overbearing father, illustrates the interplay between genetic and environmental factors in determining the course of an individual's life. For example, the relentless teasing he faced in school because of his race, his focus on arcane scientific and historical information, and his lack of athletic talent contributed to his later social difficulties. Dr. Kong also gives hope to family members of autistic children by showing that many of his earlier flaws were overcome to some extent, allowing him to integrate into society and to produce this book. Readers will follow his story with both intellectual and emotional interest.

Dr. Jessica Treisman
Professor of Molecular Genetics
Skirball Institute
NYU School of Medicine

1

Long Night's Journey into Day

A crimson sun casts long shadows across the Manhattan street grid. It is Saturday, July 26th, 2003. After a pleasant afternoon of bike riding through Central Park, I am looking forward to getting back to my girlfriend's apartment in the East Village. We meet at the Lexington Avenue subway stop on 77th Street.

Swipe. Beep. Swipe. Beep. I'm stuck at the turnstile. Jessica is upset with me. She goes on ahead through the gate (somehow her card works). Let's try a third time. Swipe. Beep. Shit. The handicapped service gate to my left is wide open and someone just walked through when the clerk wasn't looking. Should I? Damn, the train's coming.

'Excuse me,' I ask the guy at the booth, 'the machine won't read my card; can I just go through the gate?'

'Use another one!' he tells me without looking up.

I hear the train getting louder. A quick swipe. Beep. Another quick swipe. Beep. I let the woman behind me try. She gets stuck too.

'Hey, I'm telling you, none of these stupid machines work. I'm going through the gate!'

Halfway down the short flight of steps with my bike in the air above me, I hear the clerk on the loudspeaker hollering over the screeching brakes: 'You can't do that; go back up and pay! Go back up and pay!!'

'THE MACHINES DON'T WORK, ASSHOLE!' I scream back.

I can't believe I just said that. A sudden and unexpected wave of self righteous excitement, even exhilaration, washes over me. The southbound number 6 is pulling into the platform. The doors open. I glance over at Jessica; on her face is a mixed look of consternation and disappointment. So what? Was that so wrong? Now let's get this mother outta here.

Just as I'm entering the open subway doors, I feel a hand on my right shoulder. I turn around. Two uniformed officers of the NYPD.

'You didn't pay your fare,' the taller one says to me. On his badge is the name 'Garcia'. My heart starts to race and my palms sweat up. We miss the train.

'Officer, my metrocard didn't work. I tried twice on two different machines. Really.'

The shorter one, 'Cavanaugh', points towards a small crowd of people that has gathered around us. 'It seemed to work for all of them', he says.

'Well, it didn't work a minute ago when I tried to get my bike…'

'Okay, okay,' Garcia interrupts. 'You're under arrest.'

Cavanaugh grabs my left wrist and ties it to my right using a pair of plastic handcuffs. Jessica, now wide-eyed, pleads with him.

'Officer, why don't you just give him a ticket or summons? He really did try to pay. He can pay right now.'

'Are you trying to bribe us?' Cavanaugh asks, with feigned astonishment. 'Do you realize you're under arrest for jumping the turnstile?' he says, enunciating each word. 'You have the right to remain silent…'

I feel numb and powerless. This whole scene seems unreal, as if I'm watching myself in a movie. The last time I felt this way was when I was watching the twin towers go up in smoke two summers before. There's really nothing I can do, I'm under arrest, I just have to cooperate and let things be. The officers escort me up the concrete stairs. I worry about Jessica having to carry my heavy mountain bike. She gets it done. I glance nervously around at all the faces staring at us. I'm still wearing my bike helmet. What a relief. I hope no one recognizes me. I used to work at Lenox Hill Hospital, just above the subway station, until the previous summer.

There's a police van waiting on the corner. 'What about my bike, officer? I live in the East Village.'

'You can't take it with you, have your friend carry it back on the train,' Garcia says.

'What's going to happen to him?' Jessica asks.

'He'll be in holding overnight and go before the judge in the morning'.

'Officer, I can't go without my girlfriend, she has to come with me!

'Look, take your helmet off and come with us; she and the bike stay here.'

'Where's the police station?'

'69th street.'

'Jessica, meet me there!'

I want to know what time it is but I can't see my watch because my hands are tied behind by back. I pull really hard but I can get free. I look over at the digital clock in the police van. I is 9 pm. The last 15 minutes have taken forever. We arrive at the precinct station. A breathless Jessica enters with my bike a moment later.

'You made it!!' I exclaim with relief.

The station house is an old building with a high ceiling. It smells dusty and the voices echo off the walls. I think back to the first time I entered the cavernous cafeteria of the Donald Quarles School on the first day of first grade. The kids were loud and rowdy. I was really scared and started to cry uncontrollably. My teacher, Mrs. Dixon, took me to the principal's office, where Mr. McNeill gave me soothing words and shared his milk with me. Back to now. The officer at the front desk is busy directing a disheveled and disoriented man in filthy sweatshirt. He staggers out into the street. Two other officers are laughing (at him or at me?) and drinking coffee in the corner. They seem so relaxed. Is this really happening to me?

Garcia turns to the first officer. 'We picked him up at the subway platform. Turnstile jumping.'

'Okay, bring him in,' the clerk says.

I turn to Jessica. 'Take the bike home and try to get me out of here. This is a big mistake. I don't belong here. I LOVE YOU!'

'I love you too,' she says, her eyes moist.

◆   ◆   ◆

Garcia and Cavanaugh take me to a table. I am told to empty my pockets and remove my shoelaces. I do as I'm told. A pair of keys for my bike lock, my watch (Jessica's gift to me for my 34th birthday the year before), a twenty dollar bill, some loose change, my metrocard. I get to keep a quarter for a phone call. Then I am escorted to a holding cell. My handcuffs are removed. There is blood on my hands from the scratches I made trying to get those cuffs off.

There is a guy already in there. He's big and black, and slouching in the corner, his oversized unlaced sneakers propped up on the bench. I try not to act scared. Don't sit too close to him, but don't sit too far away, either. I think I'm dressed funny: a short Asian guy wearing a T-shirt and floral print shorts. Fortunately, the guy with the sneakers doesn't even glance over at me. But he seems to know Garcia.

'Yo, I gotta get my wheels back, man. I got places to go, man, shit to do. Every little thing, man, every little thing,' the sneaker guy mumbles.

'What's this, your third offense? You're driving with no license, man. What you expect? A pat on the back? Come on, get your act together, Dion.'

'I's workin' on it, know what I mean, it takes time, yo, shit don't come easy.'

'You got that right, 'specially on nights when I'm on,' Garcia says with a hearty guffaw.

'Shit' Sneakers says, then appears to fall asleep.

Garcia starts to read the sports pages in the New York Daily News. Several officers come down. One of them, a short man resembling Joe Pesci, is telling a joke. Another, with a face full of zits and looking like a high school bully I once knew, is munching on a bag of BBQ chicken wings.

'So what's the difference between OJ and Superman? Huh? Give up?'

Chicken Wings is too absorbed with his dinner to reply.

'This is the fiftieth time you're telling that stupid joke, Pigano,' Garcia says, still reading the baseball scores.

'Shut up, man, did I ask you? Anyway, what's the difference between OJ and Superman?'

'What the fuck is the difference between OJ and Superman? Hmmm, let's see, Superman is not a nigger?' ventures Chicken Wings, mocking puzzlement.

'No, man, OJ walked! OJ walked!' Pigano starts to giggle.

'You're a regular fucking Conan O'Brian, Pigano.'

I can't fall asleep. My back hurts; my ass hurts. Unfortunately, this is not a padded cell. These NYPD officers are endlessly telling crude jokes and bantering irrelevant gossip. Sneakers wakes up and asks me what I'm here for. I try to sound tough.

'Fightin' over drugs and shit, man!'

'He gives me a tired, dismissive look, and falls back to sleep. I don't think I sounded very convincing.

Finally, as I'm starting to doze off, the steel doors slide open with a screech. Pigano orders me out.

'Alright, time for fingerprints and photos,' he tells me.

After having my mug shots taken, I am led to what looks like a cross between a photocopy machine and an ATM.

'You should check this out. You're gonna love this machine! State of the art. It's like high tech. They just taught us how to use this last month. So what do you do? Are you like a computer programmer or a scientist or something?' Pigano asks me while fiddling with the buttons.

'I'm a doctor' I tell him. I shouldn't have said that, but I'm afraid that he might catch me if I lie.

'No kidding,' he says, turning to me. 'My uncle's a doctor. He's an orthopedic surgeon. He fixes bones and stuff. I kind of figured you were a doctor. You people are like really smart, right?'

'I guess,' I say, not wanting to argue with this little man.

Pigano takes my right hand and presses my fingers and palm onto a glass screen one at a time. For some reason, the computer won't register my prints. He tries at least a dozen times. I fight the urge to interrupt him. He finally calls Chicken Wings for help. Again, no success.

'Are you Chinese?' Chicken Wings asks me.

I am taken aback, but answer him. 'No, I'm Korean, why?'

'There might be something with Asian fingerprints, the ridges are too short or something. I'll get my supervisor.'

I can't believe I'm hearing this crap. I lose what respect I had for the NYPD.

The supervisor arrives. She is a middle aged woman who finally gets the machine to work. Finally! Damn idiots. I can't believe I'm relieved to have my fingerprints permanently stored in the NYPD criminal databank, but I am.

◆   ◆   ◆

I am dreaming about spending a precious Saturday evening out on the town. But that terrible metal door awakens me again. The clock says 2 am. 'All right, time to go!' an officer yells. I am half expecting someone to tell me it was just an unfortunate misunderstanding. Perhaps they're releasing me.

No chance. We are handcuffed again. Sneakers and I are escorted out into the chilly night and into a van. Two officers are sitting in the front; a third joins us in the rear compartment facing us. I look down at his gun; I wonder if he's ever had to use it. I look over at Sneakers. He's fallen back to sleep.

'Jerry, go down Lex and swing over on 42nd, let's check out the nightlife,' the officer in front of me tells the driver.

We are in Times Square surrounded by giant neon billboards for Victoria's Secret and wrap-around jumbotrons showing news from Baghdad. It is still quite crowded at this hour, but the tourists seem to have been replaced by bums and prostitutes. We slowly drive around in circles for several blocks. Dozens of yellow taxicabs pass us. The air is filled with the sounds of traffic, loud conversation, and dropping bottles. Hundreds of people walk on by without the slightest inkling of what kind of shit I've gotten myself into.

'Whew, look at the hooters on that one,' says the cop on the passenger side with a chuckle. Walking across the street in quick little steps is a black woman with very high stiletto heels and a red skintight miniskirt. 'Hey, mama, you can get arrested for hooters like that,' he tells her through the open window. She flashes a big toothy smile and blows him a kiss. 'I got your number, babydolls,' he yells back.

A few blocks away, we encounter a group of Latino men arguing on the sidewalk. The van's sirens sound, bright lights flash, and the men scatter about. We idle for several long minutes. 'That's enough fun for one night; let's drop these gents off,' says the cop in the back. We start moving downtown again.

◆   ◆   ◆

I am standing in a line just outside the basement entrance to one of the criminal justice buildings in Lower Manhattan. Ahead of me is Sneakers and about six or seven other men, all black, all handcuffed like me. Behind me is a small woman in her early twenties with tight bushy hair. She carries on an obscenity-laden conversation with herself. We are soon joined by another group of people brought in on a dark painted school bus. The line moves slowly. 'You can't keep us out here like cattle, we is cold!' a woman yells. 'That's all we is to them, animals. Wild animals,' a man replies. The cops ignore the protestations.

At the front of the line, two police clerks are cross checking papers, photos and fingerprints. One of them looks me up and down. She is a middle aged blonde woman who I imagine could once have been attractive, but now has bags and wrinkles under her eyes. 'What's with these separate prints? You got three different guys here or what?' she says in a gravely voice to the officer who has accompanied me from the uptown station. 'Yeah, we had some problems getting his prints with the new machine, so we made some duplicates.' She looks down at my file again and glances over to her colleague. 'Idiot rookie' she mutters under her breath.

The men and women are separated. Then the men are split up into smaller groups. We are led through a corridor lined with cinder blocks and exposed ceiling pipes. On the wall is a poster that says, 'You have the right to see a lawyer within 24 hours of your arrest.' That could be anytime until Sunday night. I think of all the patients scheduled to see me in my New Jersey office Monday morning. I have to be out of here. No one can know what's happening to me. Panic wells up in my stomach and chest; it is hard to catch my breath. Get a hold of yourself! Think of other things. Think of other things.

Yet another clerk behind a desk. 'Sir, when can I see my lawyer?' I ask imploringly.

'Sometime in the morning.' Relief. 'But things are slower on Sundays.'

'How much slower?'

'Might have to wait 'till Monday, depends on the number of cases.' Panic and unease fill my stomach. I'm starting to feel abdominal cramps. I hope to God I won't have to use a toilet here.

We are led past several cells, each full of lethargic people sitting around with blank stares. We are put into the last one. This cell is larger than the one in the other station. It is also much better lit by the harsh glare of numerous fluorescent bulbs. There are no windows, and no clocks. My watch has stopped working. There are about two dozen men crammed in here with me. I sit on one of the two metal benches. There is no room to lie down; another man is already sprawled out beside me. At least a dozen more are lying on the floor. In the corner, someone is sitting on the toilet taking a crap. He seems undeterred by the lack of doors or screens for privacy. In another corner is a pay phone. I reach into my pocket. Nothing. I have lost my quarter. Shit. I try but cannot fall asleep.

'Fuck this NYPD bullshit, man! It's fuckin' racism, man! Every dark-skinned brother is a fuckin' terrorist to them, man! I guarantee you, if we was white, they wouldn't bother with us, you know what I mean? How many white brothers you see in this shithouse, man?' A slight man with tan skin and a heavy Arab accent is carrying on an animated diatribe with a much taller black man in a red tracksuit and dreadlocks. 'Institutionalized racism, man, from the top down. Nine-eleven, man, they planned it all out. Who do you think was behind that, man, huh? Who do you think?' Dreadlocks just shrugs. 'The CIA, man, and Bush. You think they didn't see it coming? They ordered that shit!' The Arab's big eyes are dart back and forth wildly. This nonsense lasts several minutes before Dreadlocks retires to his bench. The Arab, still full of furious energy, paces back and forth, talking to himself. I try not to catch his eye. Too late.

'Hey, Chinaman, what they get you for?' I contemplate feigning no knowledge of English, but can't quite go through with it.

'Turn style jumping.'

'That's all, man?'

'Yeah.' I'm too exhausted to elaborate.

'That's the fucking NYPD. You see them arresting white people for things like that?'

'No,' I say wearily.

'Institutionalized racism, man!'

Some time later we are joined by half a dozen others. Among them are the first white prisoners I've seen all night. One is a very tall man in his twenties. He is wearing a faded blue denim jacket and heavily moussed black hair and looks a lot like a young John Travolta from 'Welcome Back Kotter'. Another is a balding man in his forties with a prominent paunch and rumpled business suit. The two are discussing how much time they would have to do for drunk driving. Being surrounded by these people, I am suddenly overcome by an indescribable feeling of claustrophobia and despair. I squeeze the cold steel bars and try to force them apart with all my strength. They will not budge. For a moment it becomes difficult to breath. Somehow, I pull myself together enough to approach some of the less offensive looking prisoners and ask them for a spare quarter. One of them, a man with very dark skin and an African accent, gives me fifty cents. I thank him and walk over to the payphone.

Jessica answers on the second ring.

'Sweetheart, it's me. What time is it?' I whisper.

'Half past five,' she replies with a mixture of fatigue and concern. 'Are you alright? I was so worried about what might have happened to you I couldn't sleep. Did anyone hurt you?'

'I'm alright, but you have to get me out of here, I can't take much more of this.'

'I don't know what to do!' she says, her voice cracking.

My eyes burn as I fight back tears.

'Try to get to the courthouse downtown, and find out when I can be released, call a lawyer, call your parents, call anyone, you have to get me out of this! You know this is one huge mistake.'

'I know,' she says quietly, now openly sobbing.

'I love you.'

'I love you, too.'

I quickly look around to make sure no one has been eavesdropping. I lie on the floor, close my wet eyes and pretend to fall asleep. Soon, I drift in a fitful delirium, dreaming of awakening under the sheets of a soft mattress with Jessica's naked body snuggled against mine.

Suddenly I am awake. I look down at my shorts and legs. Thankfully, no one seems to have assaulted me. But I am not with Jessica. I am on a stainless steel floor with no blanket, bathed in harsh lights and surrounded by strange sleeping men. I lie awake for a long time thinking of nothing in particular. Eventually, an officer walks in and announces it is time for breakfast.

They pass out ham and cheese sandwiches and milk. I have no appetite. I give my food to the African gentleman who lent me his quarters. He smiles at me. We tell each other our stories. He was arrested for possession of marijuana. He assures me that the judge will certainly throw out my case. I wish him luck. The despair has passed and I feel a lot better. In fact, I feel strangely exuberant. I take solace in the knowledge that somehow, this ordeal will end.

We are taken to another cell upstairs. This one is darker and older. The bars are still there, but the benches and ceilings are made of wood. A clerk calls out names one at a time. Each prisoner is told to enter a small booth and speak with the court appointed public defendant behind a window protected by thick glass. The names go by painfully slow and in no discernable order. The odor of sweat permeates the still and sullen air. Tempers flare. A tall muscular man in a white tank top who looks like he's been hardened by decades in the system paces the room yelling and cursing to himself. Everyone seems to ignore him.

'Motherfucker! You better watch out, you fucking bitch! I should have shot your fucking black ass when I had a chance, you goddammed gold-digger. After all the shit I did for you, after all we been through, you go around and stab me in the back! You take my fucking money, my fucking child support, you take ten years of my life, you shack up with other men, while I suffer in this hell. Well if you think you've seen the last of me, you better think again, bitch, 'caus you got another thing coming. As soon as I'm out I'm gonna get you, bitch! Hey get me and outta here, I wanna see my fucking lawyer, I got my fucking rights…'

'Hey shut up,' says the African softly.

'What the fuck?' responds the angry man.

'You talk too much, now shut up.'

Around noon, the lawyers break for lunch. I have not yet been called. Several hours later, the roll call resumes. 'Kong! Kong!' yells the female clerk. I wince. The cacophonous sound usually so familiar sounds oddly alien. Finally, it's my turn.

I take a seat in a low chair facing the tiny window. Facing me on the other side is a young white man in a shirt and tie. He is about my age. The man adjusts his glasses and opens a file.

'Mr. Kong?'

'That's me, sir.'

'So it says here that you were arrested at a subway station for evading the fare.' I explain myself.

'Yes, I am guilty of being obnoxious and yelling an obscenity in public. But I believe the police response was excessive and unnecessary. A ticket or summons would have been more reasonable. But, my first main concern right now is for you to please let me go; I have to see my patients tomorrow. And secondly, I really can't afford to have a criminal record.'

The lawyer listens to me attentively. 'Well, it does seem that the officers were out of line to arrest you for this little thing. My best advice to you, Mr. Kong, is to plead guilty to public disturbance, a so-called 'quality of life crime', in exchange for expunging your record in six months. I will make the recommendation to the judge.'

'What does this mean?'

'You're free to go and your police file will be destroyed in six months.'

'And no fine.'

'No fine.'

An hour later, still handcuffed and wearing filthy bike shorts and tee shirt, I am led out the door into the courtroom, conveniently located adjacent to the holding cell. There are about twenty or so people in the gallery. Jessica and I spot each other. She waves to me. I give her a big smile. The judge reads the case. 'Mr. Kong is guilty of public disturbance, and is released on his own recognizance.' The court officer removes my handcuffs. It is over before I know it. Jessica is full of joyful tears. We embrace each other tightly in the middle of the courtroom.

'I was so worried about you, I thought you would be raped,' she says quietly.

'Thank God. That was the worst experience of my life,' I say. 'Is my bike okay?'

The court officer interrupts. 'All right! Break it up! You have to clear the courtroom!'

It is three o' clock in the afternoon when I walk out into the brilliant and blinding Sunday sunshine.

2

Millennium

Anyone who has common sense will remember that the bewilderments of the eyes are of two kinds, and arise from two causes, either from coming out of the light or from going into the light, which is true of the mind's eye, quite as much as of the bodily eye; and he who remembers this when he sees anyone whose vision is perplexed and weak, will not be too ready to laugh; he will first ask whether that soul of man has come out of the brighter life, and is unable to see because he is unaccustomed to the dark, or having turned from darkness to the day is dazzled by the excess of light.

—Plato
The Republic

As the name eyes absent (eya) suggests, the eyes are missing in some eya mutants. However, stronger eya alleles cause lethality or sterility, and expression of the gene is not restricted to the eye, showing that eya has functions in addition to eye determination.

—Dr. Jessica Treisman

I rush out of the Payne Whitney Clinic of New York Hospital on a bright brisk April morning in the year 2000. I have just been diagnosed with Asperger Syndrome. There is a bounce in my step as I make my way across town trying to organize the furious rush of thoughts in my head. At last, here is a coherent explanation, a continuous thread that runs through the disparate scenes of my life, connecting the recurring themes of confusion, alienation, rejection, embarrassment, anxiety, and self-loathing. The lights of Fifth Avenue turn green, but I hardly take notice. I feel weightless, as if the psychic burden of a lifetime had been instantly lifted. By the time my head clears, I am in the Sheep's Meadow, looking towards the towers of Central Park South. A faint mist still obscuring the outlines begins to lift. In a moment of exhausted elation, I resolve to write a book about all of this.

◆   ◆   ◆

The story actually began a year earlier. On Tuesday the sixth of April 1999, I woke up, put on my thick glasses and kissed my then girlfriend Helen on the back of her pale neck as she was putting on her green scrubs. I made her take them off, and we made love. She was supposed to be getting ready for her nursing shift in the surgical intensive care unit at Mount Sinai Hospital seven blocks away. Afterwards, she gave me some affectionate words in that sweet Ulster accent I had grown to love, promising to meet me for drinks that night at the Kinsale, a local Irish bar on 96th Street. I took some comfort in this hint of short-term stability, knowing all too well that our relationship was probably doomed. I saw her out the door of her fifth floor apartment at 130 East 93rd Street. Late for work, she hailed a yellow cab on the street below. The phone rang, and I jumped. Assuming it was one of her friends (since I had no friends), and afraid of an awkward conversation, I let it ring on.

I was off that day, so I decided to walk over to a nearby coffee shop, the Muffin Man, for my usual favourite: smoked salmon and cream cheese on an everything bagel with a mug of coffee and the New York Times. Since moving back in with Helen the previous December, I had grown very used to this routine. But this time something was very different. On the first page of the Science Times, just below a pair of fascinating articles on the origin of life and the invention of writing, was an article entitled, 'A Syndrome With a Mix of Skills and Deficits,' by John O'Neil. Intrigued and with some time to spare, I read on. It was about odd self-absorbed children often with above average IQs who suffer from social retardation. The hair stood up on the back of my neck; I was floored. This article was about me.

Remarkably, this was the first I had heard of 'Asperger Syndrome'. It had not come up in four years in medical school (including courses in psychology, psychiatry, and neuroscience) or in my three years as a resident in internal medicine. I was vaguely familiar with its cousin disorder, autism, but never suspected that I might be afflicted with something like it. I was hungry for more information. That night, I met Helen for bangers and mash and a pint of Guinness. I did not mention Asperger Syndrome, but it was really the only thing on my mind.

Over the next three months I scoured the medical literature. Surprisingly little had been written and even less was known about it. The best books on the subject were written by a pair of British psychologists, Uta Frith (**Autism: Explaining the Enigma** and **Autism and Asperger Syndrome**) and Simon Baron-

Cohen (**Mindblindness: An Essay on Autism and Theory of Mind**). But the more I read, the more I became convinced that I, too, was suffering from this high-functioning variant of autism. It explained everything: the social ineptitude, clumsiness, inflexibility, impulsivity, naivety, love of routine, difficulty getting others' point of view, the self-centeredness, self-righteousness, social anxiety, and so much more besides.

Unfortunately, it was during these very months that my relationship with Helen took its final tailspin. She was in love with me, or so she said, but for two years she had wanted a commitment to marriage that I was unprepared to give. If only she would be willing to go out with me indefinitely as my girlfriend, no strings attached. To make matters more complicated and painful for me, my parents had stopped talking to me for cavorting with a 'white devil'. At one point a year earlier, I even proposed to Helen on impulse, and was prepared to weather whatever storm my parents might dish out; but the continuing stress made me change my mind. Caught between the narrow minded bigotry of my parents and the aggressive possessiveness of my girlfriend, I very nearly suffered a nervous breakdown. In one particularly nasty but memorable tiff at the Muffin Man, Helen smashed a bagel and cream cheese onto my face and dumped a cup of ice coffee on my head in the middle of the morning rush, much to the shock and embarrassed amusement of several dozen customers (we made up by having wild sex on her bathroom floor an hour afterwards). But after a series of false starts, we finally broke up at the end of June.

In the autumn and winter of 1999 I was working part time for an endocrinologist. His office was in the Majestic Building just across the street from the Dakota, where John Lennon had been shot 19 years before. Every morning, I would take the delightful walk across the Park from my apartment on the Upper East Side. It felt good to be young and free and single in New York City. But not everyone approved. My parents wanted me to move back to Jersey and start my own private practice. 'Where's the future for you there? You're throwing away $1800 a month renting a tiny studio apartment,' they pointed out.

It slowly dawned on me that I was not, in fact, getting the best deal from my boss. As I was working full time hours for part time wages, I asked him when I could expect a raise or a full partnership. 'Well, I'll have to give that some thought, but I really appreciate your support,' was the standard reply. I felt genuinely reassured by lines like that. I was sure he had my best interests in mind. 'That's what he said, so I'm sure he means it,' I kept telling my skeptical parents.

It took nearly a year for me to realize that his best interests were his own, and another six months for me to get myself together enough to quit.

Being an underemployed doctor gave me the time to pursue other interests. I signed up for an internet dating service geared towards graduates of the Ivy League and medical schools. My bio went like this: 'Hello, I am a 31 year old physician living and working in Manhattan. I graduated from MIT '90 (biology), University College Oxford '91 (British History), & UMDNJ Medical School '95. My hobbies are reading, traveling, and bike riding. I am also an incurable Anglophile. If you're interested, we should get together.' Hardly any women responded, but I still managed to set up about a dozen first dates. The nervous anticipation was addictive. Unfortunately, the excitement invariably ended in disappointment for me. All too often, the women were too old, obese, or just plain unattractive. I was endlessly fantasizing about sexy blond 'girls next door' from porno centerfolds. On those few occasions when I found an attractive woman, I never got a second date. It was puzzling. I went to the bookstore and bought manuals on how to pick up or date women, but to no avail.

In March, 2000, I came across a profile that I thought would be a sure thing. The young woman in question was a 25 year old Korean American in her third year at Harvard Law School. Although I am generally not physically attracted to East Asian women, her photo was appealing enough. I knew that my parents would approve of her background. Also, I expected her to be an intellectual match for me. How could I possibly go wrong here?

I planned the meeting meticulously, starting with a reservation at the Cambridge Marriott and a box of Godiva truffles. I would drive up to Boston on Saturday morning, meet her outside her dorm, take her out to dinner and a comedy club, and, if all went well, invite her for brunch at the Massachusetts Institute of Technology (my alma mater) on Sunday. Everything went according to plan. I was the perfect gentleman, opening doors, complimenting her on her dress and appearance, and picking up the tabs. We had cocktails and dinner at the Border Café in Harvard Square. I had a wonderful time, talking about myself, Korean political history, neuroscience, and the creation of the universe, among many other things. We lingered for hours, skipping the comedy club. I felt confident that I really wowed her with the wealth and breadth of my knowledge and intellect. How could she not be totally impressed? I was finally about to capture a woman my parents would be happy to accept.

The next morning, I walked up Memorial Drive beside the frozen Charles River. When I knocked on her door, she didn't respond. Impatient, I knocked again. Hadn't she agreed that I could come over when I called her an hour earlier? Then she appeared in her bathrobe. It seemed a bit forward, and I was getting excited.

'I'm sorry, Henry, I don't feel too well this morning,' she said.

'Oh,' I said, taken aback. 'You seemed okay last night.'

'Yeah, I think I'm coming down with something.'

'Well, uh, maybe if you feel better this afternoon we can, uh, still hang out,' I said.

'No, I really have a lot of work to prepare for. I think I should rest today. But thanks.'

A wave of panic and indignation suddenly rushed over me. Why was this was happening?

'I got you this box of chocolates,' I blurted, thrusting the box into her chest.

'No, you really shouldn't.'

'Okay. Well, I guess I'll see you around.'

I walked back to my car and drove straight back to my parents' house in a snowstorm.

I emailed her several times over the next few weeks, but she never responded.

That spring, I saw a psychologist. He was a stuffy looking guy in his late fifties. There were lots of books on psychoanalysis lying about his dark and cluttered office. I asked him if he thought I had Asperger Syndrome. The man didn't know what I was talking about. After I explained it to him, he reassured me that there was nothing serious to worry about.

'You're a bright guy; you'll be very successful. There were some difficult times in your childhood that left negative imprints on your personality, but there's no reason why you can't learn to adapt. You certainly are not autistic,' I was told.

He charged me $200 for that malarkey.

A few weeks later, I saw a psychiatrist, a friendly chap whom I knew from my hospital. His office was bright and neat. There were lots of books on psychopharmacology on his shelves. I didn't mention Asperger Syndrome. He listened patiently and seemed less judgmental than the psychoanalysis guy. At the end of the interview I asked him what he thought.

'Well, there's certainly a lot there to work with. The traits you talk about may suggest something on the autistic spectrum, but you'll need more specialized assessment. To be classified as having a disorder, you really need to have difficulty in everyday life. You seem pretty well compensated,' he reassured me. 'But I can prescribe some anti-depressants.'

This was an improvement, but not exactly what I was hoping for.

'Not really. I think I need help,' I told him.

'I can refer you to a developmental psychiatrist named Anne McBride. She specializes in Asperger Syndrome.'

◆   ◆   ◆

I am waiting to meet the doctor in her waiting room at the New York Hospital. There is a little boy sitting on the carpet, intently stacking plastic blocks on top of one another. His mother looks over at me without smiling. A few minutes later, Dr. McBride emerges in her white coat. She is a slight woman in her forties with reddish hair and large round glasses. She ushers me in.

'So tell me about yourself,' she says, reading over my intake assessment.

Shaking with excitement, I take a deep breath, and take off:

'Well, I read this article about Asperger Syndrome last year and I think I have it. I have a hard time understanding what other people are thinking, and it gets me into embarrassing and sometimes inappropriate situations. I trust people too much, and that gets me hurt a lot. A lot of times, I just smile and pretend I know what they're talking about when I really have no idea. Yesterday, I was having dinner with some colleagues from work and when they started some office gossip, the conversation kind of flew by me, like they switched into a foreign language. It was really uncomfortable. I had to smile so much, I was sweating.

'I don't have many friends. I have acquaintances at work, but few real friends. I had a good friend in college, but we lost touch when I went to med school. I had a friend in med school, but we lost touch when I started my internship, and so on. I would like to make more friends, but I get nervous. I just try to pretend I'm 'normal'. But that gets really hard sometimes. People might think I'm weird because I don't have any friends, so I don't really try to approach them. But if I don't approach them, they can't become new friends. It's a big 'catch 22', you know?

Dr. McBride keeps writing.

'I have a hard time meeting girls. I was a virgin until I was 28 years old. I'm not gay or anything. Don't get me wrong; girls really turn me on. I like to watch porn and stuff. But I don't know what to say when I go to bars. I did meet my first and last girlfriend at an Irish bar four years ago, but that might have been an accident. It still amazes me how I pulled it off. But I had to break it off because my parents disowned me for going out with a white girl. Lately, I've been going on lots of first dates, but I don't get any second dates. I don't know why. I'm well educated, reasonably good looking, and try to be really nice to them, but there might be something about me they're not telling me. When I see all those Wall Street assholes going out with beautiful girls, it makes me really angry. I think it's so unfair.

'I have a really hard time organizing myself. It's hard to figure out what to do with my free time, unless someone tells me. It's hard for me to start a new project and hard for me to stop something once I start it. It's like I have too much 'inertia'. I'm pretty good at following directions, like I can make dinner when someone tells me what to do, but I can't think up too many new things, and I can't do more than one thing at a time. Also, I can be pretty funny when I'm comfortable and someone gets me going, but not when I'm alone with someone attractive.

'I find myself thinking about weird things almost all the time. Lately it's been about Asperger Syndrome, but before that I was really into British History, and before that it was science and science fiction. I've memorized lots of important dates in history, like battles and the reigns of kings and things like that. It's sort of effortless. Sometimes, I spend hours thinking about some obscure dates and places in history or writing up lists of the most important British scientists, or Wimbledon champions, or whatever.

'I'm really bad at sports. I got teased a lot in school for that. Kids used to call me names. But they called me names for lots of other reasons, too. Like 'four-eyes' because I have thick glasses, and 'chink' because I'm Korean, and 'geek' because I like to read the encyclopedia, and 'cry-baby' because I was always crying for my mother.'

'Tell me about your childhood,' Anne asks.

'I was born in Seoul, South Korea in 1968. My parents and I immigrated to America in the summer of 1973. My dad is a doctor and my mom is a housewife. They still have a hard time with English.

'I am in excellent health, have no allergies, don't take any medications, and don't smoke. I do enjoy drinking alcohol regularly. I have a family history of heart disease on both sides. Also, there's some history of ill-defined mental illness in my dad's side of the family. My only surgery was for a lazy right eye at age six. I am naturally left handed, but my maternal grandparents forced me to learn to use my right had for eating and writing. I didn't stop breast feeding until age four and didn't stop wetting the bed until age six.

'My brother was born in 1975. The next year, my dad joined the US Air Force and we were stationed all over the place. I lived in Charleston, South Carolina,

Biloxi, Mississippi, and Lompoc, California, before coming back to New Jersey. I liked moving around and seeing the country, but I didn't like having to start in new schools every year.

'In South Carolina, my dad's father moved in with us. That was terrible because my dad and my grandfather got together and beat my mom on a regular basis. My dad also used to beat me up for not doing well in school and stuff like that. On top of that, I was bullied by the other kids in school, especially since we moved every year. It was traumatic.

'Things got really bad in Junior High. I was teased so much, my parents had to move me out of the public schools. But the Catholic school wasn't much better. It wasn't just the harassment. I just felt really alienated. It's hard to describe. High school was somewhat better. At least by then, I had something to focus on: academics and getting into college. But I didn't have many friends, didn't go out with girls, and never went to a party.

'I thought MIT might be better. At last, I was going to a school full of nerds and chinks like me. But I didn't feel at home with the nerds and chinks either. They would go to geeky parties and have sex with other nerds and chinks. I still felt really different. I just concentrated on my studies. I had to get to med school.

'Between college and med school, I spent a year hanging out in Europe. That was a really fun year for me. But I still didn't know how to connect with women in a meaningful way. After that, I became totally obsessed with beautiful girls with upper class British accents.

'I was in med school between 1991 and 1995. I learned everything about female anatomy and physiology, delivering babies, HIV and STD, and various aspect of human sexuality. But it was still all theory and no practice.

'I was a medical intern and resident up the road at Lenox Hill Hospital between '95 and '98. That's when everything finally changed for me. I met Helen, my first girlfriend, at a bar just before Halloween 1996. We broke up last year, just after I learned about Asperger Syndrome.'

I look at my watch. The hour is already up. I am still shaking, but no longer nervous.

'So, Dr. McBride, what do you think?'

'I think you may have Asperger Syndrome.'

I can't help but grin. 'Wow. What makes you think that?'

'Well, for one thing, when you got excited and started on your monologue, you lost eye contact with me.'

Catharsis. I feel extremely good and satisfied, almost post-coital. Anne McBride may not realize it, but this is one of the greatest moments of my life, right up there with my first orgasm with Helen and getting that big fat acceptance letter from MIT.

◆   ◆   ◆

It is early May, 2000. For the first time in my life, I'm actually earning some substantial money. After all the crap with which I've had to put up, I think I deserve it. Perhaps I'll put a chunk of it in the booming stock market. Of course most of it has someplace to go: malpractice insurance, Manhattan rent ($1750 for a studio), psychotherapy sessions ($200 an hour). But there is one last expense I've been brooding over for months: laser eye surgery.

Dr. Friedman assures me that it's safe and simple. 'Trust me. You'll be happy. The girls are going to like you.'

I am worried about possible problems with night vision and long term side effects. An ophthalmologist friend of mine (who, like Dr. Friedman, is comfortable with his glasses) recently warned me that I might go blind. I think he might have been half kidding. I'll take my chances. I've had severe myopia and astigmatism since age six, and often fantasized about waking up and being able to read the clock on the wall. To see the world as it was meant to be seen, and to be seen by the world as normal; how could that not be wonderful?

I didn't always have glasses. There is a photograph of me as a four year old standing shoulder high in a lavender and ochre flower field somewhere in Korea. I am not wearing glasses. But if you look closely, you can see that there is something not quite right with me. My eyes are crossed. I had a defect known as 'strabismus'. It is usually caused by a defective nerve or weakened muscle responsible for moving the two eyes together. Congenital strabismus is often harmless aside from its social stigma. But it can lead to permanent diplopia (double vision) if corrected too late. It was for this reason that I had surgery for my 'lazy' right eye at age five. A few months after this, I was found to be myopic (nearsighted). I'm told that my strabismus was unrelated to my myopia, and furthermore, that both defects are unrelated to my Aspergers. I seriously wonder about this.

I find myself lying back on the surgical table in the laser suite, trying to relax. My eyes have been taped open.

'All right, you're going to feel some pressure. Just pressure, no pain, okay?'

I feel a lot of pressure. And, yes, it is starting to hurt.

'See the blinking red light directly above you? Just keep looking at it.'

I stare at the light, but the pain is getting quite bad. Everything becomes a blur, and then I lose my sight.

'You're gonna be okay. Just a few more seconds.'

I hear a high pitched series of beeps and smell something burning. The laser is carving my eyeball.

It's over in less than thirty seconds. The other eye isn't so bad. Twenty four hours later, my right eye is nearly 20/20 and my left about 20/30. My LASIK surgery is complete exactly a quarter century after my strabismus correction.

◆   ◆   ◆

August 2000. I come across Jessica's bio on the 'Right Stuff Dating' list: '36 year old, Oxford BS '85, Rockefeller Univ PhD '91, biologist; attractive, successful, intelligent, and ready for serious relationship.' I email her.

She replies the following week:

'Thanks for the email. It seems that we have some things in common, like both being alumni of University College, Oxford. Perhaps we can meet for a drink later this week.'

Our first date is at the Royalton Hotel Bar on 44 West 44th Street. It turns out that she is a molecular biologist at NYU working on, among other things, the genetic details of fruit fly eye development. She is extremely intelligent (having scored a perfect 1600 on her SATs and learnt to read before she could walk) and well educated, with an English accent, all qualities I find very attractive in women. She is also very nearsighted and terrible at sports. Finally, she comes from a family of prominent psychologists, artists, and social scientists. Her mother is a world famous cognitive psychologist whose work on attention and consciousness I have admired since medical school.

Jessica and I have some traits in common. But we are also quite different. She is politically liberal. I am less so. She loves to read fiction. I prefer nonfiction. She understands the way people think, even if she would rather watch and listen. I am generally socially clueless, but try to participate when I can.

This millennial year is turning out to be the best in my life. Meeting Jessica, having eye surgery, and getting the Asperger diagnosis are extremely significant. I like to think they signify the end of my blind wandering and silent suffering. It seems that I am starting to overcome ignorance, myopia, and mindblindness and 'seeing' the world in whole new ways. The quest for focus and clarity is what this journey from autism to understanding is all about. I will have to write a book about it.

3

Klutz

'Don't pick Henry! He can't even see the ball. Get Warren.'

'No, Warren's too fat!'

'Yeah, but at least he can kick!'

It is the winter of 1976. The kids in my third grade gym class are picking teams for kickball. Warren (the fat one) and I are the last ones standing out. Out of pity, Coach Floyd randomly assigns us to opposite teams.

'No, coach, we don't want Henry on our team! He's gonna make us lose!'

'Come on guys, play fair. Everyone gets a chance,' the coach says.

Our team is first up. Slim pitches the ball to Frankie. The kicker runs up and pounds it over left field. He quickly scrambles to second base. Frankie's twin brother, Baby Louie, is up next. Louie kicks it into foul territory.

'You guys suck!' someone on the other team jeers.

'Oh, yeah? C'mon you faggot, you can't even pitch!' Baby Louie retaliates.

The next pitch is good. Baby Louie smacks it right up the middle and into the parking lot for a homer.

'Hey, douche bag! Who sucks now?' he yells crossing home plate.

By the time I'm up, the score is 4-0 with two outs. Slim tells everyone on his team to come closer.

He rolls the ball towards me. I run up, but misjudge the speed and position of the ball. My foot meets thin air. I hear giggling in the background.

'Strrriiikkkee one!'

The second pitch is a bad one, bouncing off the ground. I probably shouldn't go for it, but can't help myself. More air.

'Strrrikkkee two!'

'Come on, Henry, look at the ball!' screams Frankie.

The third pitch is better. I run up, but suddenly realize my left foot is too close to the ball. I try to compensate by using my other foot. Too late. My awkward kick sends the ball straight towards the shortstop. He easily intercepts it and slams it into my shoulder before I'm half way to first base.

'You're out!'

My second time at the plate, I manage a decent kick and make it to first base. A little later I'm running towards second as fast as I can following another good kick by Baby Louie. Then suddenly, Corina, a girl who's always picking on me, sticks her foot out. I land hard into the dirt. My head shakes violently and I feel dizzy. Meanwhile, someone tags me out.

We lose the game 11-10. Frankie and Baby Louie tell me I'm 'nothing but a god dammed klutz' who should never be allowed to play kickball again. Thank God it's over.

By the time I run home, the tears have stopped coming out of my eyes. I feel a lot better. Waiting for me is the broken TV set I made my parents drag into the apartment from the dumpster the night before. As soon as I'm finished with dinner, I take my screwdriver and pliers and rush to work. Out come vacuum tubes, circuit boards with little capacitors and resistors welded onto them, and lots of multicolored wires. I stack them neatly on the living room floor. It makes me feel really good inside to take things apart.

The next day is really cold. Frankie, Slim, and Baby Louie invite me to walk home with them.

'My mom is coming to pick me up,' I say.

'Just walk with us for a little bit. We found something really neat,' Frankie tells me.

'Well, okay, I guess.'

Outside the school fence is a little lake. It is frozen. The three boys race onto the ice and start sliding around.

'Hey, this is really neat! Let's play hockey!' Frankie exclaims, pulling out a plastic puck out of his jacket pocket.

'I don't know, my mom's gonna be coming soon,' I say uneasily.

'Henry, why don't you be the goalie?' Slim asks me pointing to a hole in the ice. 'Just stand out there. If the puck goes into the hole we win.'

'I think the ice is kind of thin,' I say, gently walking over.

'Naa, you'll be okay. Now get ready.'

Just then I hear a voice yelling in Korean. It's my mother.

'Get away from there! Are you crazy?'

And to the other boys in broken English: 'You go away! You! You! You go home now! I call your mother!'

They disperse laughing.

I don't play with the other kids any more. I don't go out at recess. Someone calls in the school psychologist to evaluate me. She asks me some questions to see if I understand English. Then she gives me a kind of IQ test. On the last part of the test I am given a bunch of colored blocks and told to rearrange them into composite shapes. She tells me I did really well on that part, but doesn't tell me what my IQ is.

◆   ◆   ◆

There are numerous reports of sensory deficits in autistic individuals. Some of them may be overly sensitive to noises, bright lights, odors, and textures. Autistic children are known to throw tantrums in busy public places like shopping malls. But is it because they actually perceive things differently from '**neurotypicals**' ('normal' people)? Does a siren sound different, a flashlight look different, a perfume smell different, a shirt feel different when heard, seen, smelled, or worn by an autistic individual? This is a very difficult question to answer because you can't objectively put yourself into another person's mind. Is the red I see in the rose in front of us just as 'red' for you as it is for me? Does it smell as sweet? Do the thorns feel as sharp? We would probably agree that the color sensation is very similar, but the feelings and thoughts we get are not. But even if both of us say we see the same intensity of red, and feel the same discomfort from the spiky thorns, how can we objectively compare the quality of the sensations?

In philosophical circles, this is known as the 'Qualia problem'. It is closely tied to two other questions: what is it like to be someone (or something) else, and what is the nature of the subjective self? These lie at the heart of science's next Holy Grail: Consciousness (see **Appendix 2**).

Let's go back to the issue of purported sensory/perceptual abnormalities associated with autism. Some studies report visual or tactile deficits or auditory hypersensitivity in certain populations of autistic children; however, the majority of autistic individuals perform normally on simple tests of visual or auditory acuity. Autistic children have been found to perform unusually well on tests of 'local coherence' such as the **hidden figures test**. This test involves picking out simple geometrical shapes from a larger composite picture. The problem seems to lie in higher order perceptual integration. Children with autism or Asperger syndrome seem to have a loss of 'central coherence.' They miss the forest for the trees.

In order to see how those on the autistic spectrum experience the world, we first need to understand how the normal brain works. Scientists have a rather good understanding of at least the first stages of perception. We know that the senses of touch, balance, hearing, smell, sight, and taste are mediated by receptors in the skin and joints, in structures deep inside the ear and nose, in the retina at the back of the eye, and in taste buds on the tongue. These receptors convert or 'transduce' phenomena in the outside world like heat, gravitational acceleration, odor molecules, and waves of sound and light into patterns of electrical activity (called action potentials) in the sensory nerves. Perception occurs as the information conveyed by these patterns is relayed to higher and higher processing centers in the central nervous system. I will take you on a brief tour of the sensory system scientists know the most about: vision.

I am looking at a reproduction of Seurat's famous pointillist picture of a picnic in the park. Photons of light from the window are reflected off the surface of the painting and enter the lenses of my eyes. Each lens flips the image upside down and focuses it (more or less) on a thin curved sheet of retinal cells. There, specialized receptor nerve cells (the rods and cones) contain chemicals that change conformation when exposed to the energetic photons of light. These chemical reactions, in turn, produce changes in the electrical potential of the rods and cones that are transmitted to other neurons connected to them. If sufficiently stimulated, these second neurons produce electric spikes (action potentials) that spread down long output appendages (axons) connected to input appendages (dendrites) on still other neurons. The activities of thousands of nerve cells containing a sort of 'neural code' are channeled into multiple bundles of axonal wire continuously streaming from the two eyes through the optic nerves. These nerves eventually arrive at the primary visual cortex. The perceived visual image that was initially projected on a tiny sheet of cells on the surface of the retina is now transferred to another sheet of cells on the back of the brain.

The Seurat painting we 'see out there' has now been deconstructed into a partially processed buzz of electrochemical activity spread out over millions of nerve cells and their billions of connections, called synapses, scintillating with the minute bursts of millions of action potentials. Differentiated and deconstructed as the sensory input may now be, there is order underneath the chaos.

The visual field is 'mapped' point by point onto the surface of the visual cortex. The **patterns** of neural activity **code for** simple features like edges, movements, and colors in particular parts of the visual field. Specific columns of neurons in V1 are dedicated to analyzing the input from each eye, specific colors, or certain orientations of edges. Scientists can now visualize, in full breathtaking color, the activity of the cortical columns in living cats or monkeys by a technique called optical imaging.

The next task for the brain is to take the differentiated pieces of sensory input (now encoded in V1) and integrate them into something coherent—a 'bigger picture' if you will. Something very interesting happens further 'upstream' of the primary visual cortex. Back in the late 1970s, neuroscientists found that the processing of visual input is divided into two channels coming out of V1. They are the **dorsal stream** (so called because its nerve tracts run along the top, or dorsal, part of the brain) and the **ventral stream** (whose fibers run along the lower, or ventral, part of the brain). The dorsal stream specializes in the location of objects in space, allowing one to perform such tasks as grasping a moving branch or catching a football pass, while the ventral stream deals preferentially with the recognition of objects. Thus the dorsal system has come to be known as the '**where**' or '**how to**' pathway and the ventral system as the '**what**' pathway (**figure 2**). Brain injuries can cause specific deficits in either the processing of motion (such as perceiving the flow of running water) or in the perception of form (such as recognizing simple geometric shapes). Many such patients have anatomic lesions in either the dorsal or ventral systems respectively.

The other sensory systems are similarly designed. Sounds, touch, temperature, tastes, and smells are all deconstructed and then recreated in their respective areas of the brain. A different area of cortex is allocated for each modality. The differentiation is followed by integration in the association areas of the cortex. Like a fine motion picture that somehow suspends our disbelief for a few hours, the synchronized activity arriving at these areas acts to 'bind' the disparate properties of sensory perception into a seemingly coherent and ongoing narrative.

The products of our perceptions are working approximations of external reality, just as a pointillist painting, when viewed from afar, approximates a real scene. But in the end, it is all just an elaborate illusion. Seurat's painting is made of oil paints on canvas, not grass and water and flesh. Likewise, the vision of Seurat's painting in my mind's eye is not made of oil paints, but of complex ensembles of neurons firing tens of thousands of action potentials per second from my retina to my visual cortex and beyond.

The sights and sounds we see and hear really do seem to emanate from things 'out there', rather than from stuff going on in our heads. The simple explanation for that is that evolution by Natural Selection has made our mental approximations of external reality match very tightly to the physical world, at least in those dimensions that are important for our survival. The view directly down from my window above the East Village really does appear to be approximately 150 feet above the East Village. If it did not, perhaps due to some broken gene affecting how my brain perceived distance or perspective, I might be prone to step out into the abyss, forfeiting my chance of passing on any such 'altitude perception gene' to my progeny. On the other hand, equally 'real' phenomena like ultraviolet light, magnetic field intensity, or relativistic time dilation are inaccessible to our naked senses. This is because evolution has neither felt the pressure to select, nor had the biological substrate with which to create the means for sensing them in our species.

Our perception of a coherent reality is contingent on complex and creative neural ensembles which simultaneously deconstruct and integrate information representing different aspects of sensory input. This process can be interrupted at multiple levels. Lower level interruptions, say at the sensory receptor stage, can cause deficits like congenital red-green colorblindness or conductive hearing loss. Damage further upstream in the association areas of the brain can lead to more specialized deficits. Examples include **optic ataxia** (the inability to perceive motion) due to lesions in the dorsal visual stream, **apperceptive agnosia** (the inability to perceive shapes) from lesions to the ventral visual stream, and even **prosopagnosia** (the inability to perceive faces) from damage to the **fusiform gyrus.**

I believe that the sensory/perceptual problems in classic autism (the hypersensitivities and irritability due to 'sensory overload') are not lower level or 'downstream' defects affecting the sense organs or the primary sensory cortex. Rather, they are a result of deficits much further upstream, in those areas of the brain responsible for high level integration of sensory input.

The notion that autism preferentially disrupts higher level perception while leaving the lower levels intact, thereby knocking out the ability to 'get the gist' while preserving and perhaps even enhancing the tendency to notice the details is an attractive one. If this is true, then what the autistic person sees in the Seurat painting is less a scene in the park than a near random pattern of colored pixels. It is as if a neurotypical viewer were seeing a work of abstract expressionism by Klee or Kandinsky instead.

◆   ◆   ◆

It is 1983, my sophomore year in High School. The sixth period bell rings; time for Phys Ed. Shit, we're going to be playing volleyball again. I got so nervous this morning just thinking about it that I had diarrhea. If only I could get a doctor's note to skip class, as Tim and Phil always manage to do, it would be such a relief. I retreat into an empty corner of the boys' locker room and quickly change out of my school uniform and into the regulation gym shorts and shirt. There is loud yelling and laughing coming from the other side of the room. The boys' varsity soccer team is getting ready for their away game. They are excused from gym class. At least I won't get picked on too badly.

I walk out onto the polished wooden floor. The basketball backboards have been cleared up to make way for the volleyball net. Voices echo off the walls. Everything seems so imposing. Carrie, Wendy, Elena, and the rest of the popular girls are hanging out in the back. A few of them look over at me and the other boys and start giggling. I hope I don't embarrass myself again.

'THINK QUICK!' someone yells out.

I turn around. A volleyball hits me on the head and my glasses fall off. I hear laughter. I quickly pick up my glasses and put them back on, pretending nothing happened. I feel really confused and upset. Where the hell is the coach?

Just then, Coach Hudack leisurely saunters in. Everyone shuts up. He whispers something to Wendy; the two of them laugh.

George and Carrie are on my team. George should be playing soccer this year, but is sitting out for bad grades. He looks over at me with a menacing stare.

'Kong, when the ball comes towards you, just try to block it, that's all, no fancy set ups or anything. Just do what I tell you. Use your brain! Got that?' he tells me.

'Yeah.'

'Good,' he says making a thumbs up sign. He looks over at Carrie shakes his head and rolls his eyes.

The game starts. I assume my usual preparatory stance: two flat feet spaced widely apart, stubby legs bent slightly at the knees, head tilted back, eyes squinting through thick glasses that keep sliding down my nose, arms and fingers outstretched above as if to shield me from the ball.

Most of the time, the ball comes far enough from me that someone else hits it. Relief. I manage to keep a few balls in play and even serve over the net once. More relief. I look over at the clock. Still another 15 minutes to endure.

The score is tied. Richard, the other team's captain, serves. The ball slowly arcs over the net and straight towards me. Panic. I run the routine over and over in my mind: either wait for the ball and set up if it's high, or bump it underhand if it's low…set up if it's high, or bump it underhand if it's low…set up if it's high, or bump it underhand if it's low. But this one looks like it might go out; I hope it goes out. I do nothing. No, it's coming in! My arms are still way over my head. It's too late to hit it overhead. I freeze. I see George to my left as the ball comes towards the ground in slow motion. He is staring right at me, his eyes bugging out. I have to do something! All I can muster is a lame upward two handed thrust *after* the ball bounces six inches from my foot.

It all comes to me now. I should have adjusted a bit to my right and hit the ball underhanded. George stomps over to me, his huge chest heaving, his long blond hair and facial stubble barely concealing a beet red face. He is very angry. I break out in a cold sweat.

'HEY GEEK! GET WITH THE PROGRAM!' he yells at the top of his voice. I look around. Everyone on the other team starts to laugh. Coach Hudack is nowhere in sight. Neither is Wendy. I wish I could just disappear too.

Just then, someone comes up behind me and yanks my shorts down. I quickly pull them back up. I look around. Carrie and Elena cover their mouths, but I can see they are laughing too.

◆   ◆   ◆

Imagine any voluntary activity: answering the phone, brushing your teeth, playing the piano, catching a football pass, grabbing a cold beer from the fridge, driving down a rain-slicked road at night. We think of them as single acts. They are not. Each is a complex composite made up of little movements like reaching, pulling, and twisting in a specific, finely-coordinated sequence. Each little movement is carried out by muscles under the command of their corresponding motor nerves from the spinal cord. But voluntary movements are more than simply the sum of their parts. When we walk into the kitchen to get that beer, we experience a unitary, coherent action, rather than dozens or hundreds of hand, finger, and eye movements and thousands of muscle contractions that go into a successful act of beer retrieval. What we are conscious of is a high level 'action schema'. Just as in sensory perception, where the mind's eye sees an integrated representation of the real world rather than thousands of little components, the motor system gives the brain an integrated representation of motor activity.

If you think about it, all voluntary actions share a common theme: incoming sensory perception linked to outgoing motor activity. It is rather like the classic knee-jerk reflex, but at a higher level. To a wide receiver in football, an incoming pass is much more than a spiraling two-dimensional oval in space. It becomes part of a 'play' involving the positions, trajectories, identities, and motives of the other players relative to the receiver, the field, the football, the time clock, and the score of the game. At this juncture, both perceptual and motor representations are highly abstract. As discussed earlier, sensory perceptions are bound into something coherent—a creative facsimile of the real world. It is upon this facsimile that voluntary motor plans are constructed.

Anatomically, the sensory-motor loop involves the linkage between the parts of the brain that contain the high order representations of sensory perception (the association areas of the sensory cortex) and the parts of the brain that contain the high order representations of motor preparation (the **supplementary** and **premotor cortex**, along with the **cerebellum**, and the **basal ganglia**). Information in all of these parts of the brain is organized 'somatotopically'. That is, each part of the body is 'mapped' onto a corresponding part of the brain. Certain parts of the body, such as the face and fingers, receive much more attention in both the sensory and motor maps. This is why the sensory and motor 'homunculi' are so funny looking. Evolutionarily, this makes perfect sense. We use our fingers and mouths to interact with the world a lot more than, say, our backs. An individual who, for some genetic reason, had slightly more sensitive lips or slightly more dexterous fingers might have had a slightly better chance of getting a female to have sex with him. The genes that build brains with slightly better face and hand representation became more prevalent in the population, eventually resulting in the brains we have today.

Somatotopic organization exists in the primary sensory and motor areas, the higher order cortex, and even in the cerebellum and the basal ganglia. The cerebellum, in fact, has three separate body maps. It is clear that action at a more general, schematic level is coded in these latter maps. Activity in the higher order motor cortex represents the earlier stages of motion preparation (thinking about walking into the kitchen), while activity in the primary motor cortex, the cerebellum, and the basal ganglia represents the fine-tuning of actions in real-time (opening the refrigerator door, reaching for the bottle of Corona).

The basal ganglia and the cerebellum form two distinct modulatory loops within the larger sensory-motor loop. The basal ganglia allow the individual to choose between acts, moods, and thoughts. The cerebellum is a fine tuning machine that 'filters' sensory-motor activity to produce smoother, better synchronized motions (and possibly thoughts as well). These two structures working through the important relay station of the thalamus, help bind the perceptual stream with the motor stream **(figure 1)**.

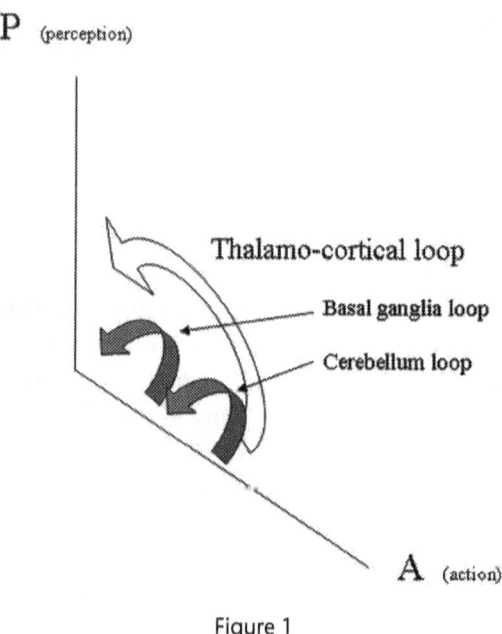

Figure 1

Where does perception end and action begin? There is no single answer. Rather, there are multiple connections between the two pathways starting as low as simple reflex arcs in the spinal cord. There are also direct connections between the primary sensory and primary motor areas (via the thalamus), and indirect ones through the cerebellum, and basal ganglia. But perhaps the most elaborate link between the realms of perception and motion involves the **superior temporal cortex** (STC). This is the final common destination for the dorsal and ventral visual streams **(figure 2)**.

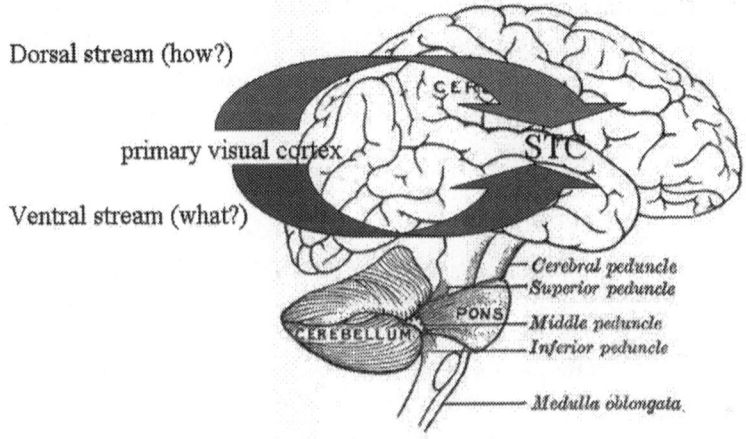

Figure 2: the superior temporal cortex

The dorsal stream (the 'how to' pathway) processes higher visual perception for the position and trajectory of objects **relative to the observer**. Presumably, this pathway evolved so that monkeys would be better able to reach and grasp swaying branches and fruits. Damage at various points along the dorsal stream can cause several types of deficits including optic ataxia (the inability to physically manipulate objects such as inserting and turning a key) and motion agnosia (the inability to perceive motion). All of these defects are **egocentric**; they involve the misperception of objects relative to the observer, which results in defective visually guided action. The ventral stream (the 'what' pathway) processes visual perception for the features of objects, including distinguishing such complex ones as faces, animals, tools, and landmarks. Damage to the ventral stream produces misperception of objects independent of their position. The ventral stream is therefore **object-centric**.

The two streams converge in the STC, which also receives input from the other sensory areas (the auditory and the somatosensory cortices). All this highly processed information is used to construct a coherent sense of space surrounding and enveloping the observer. In addition, salient things and beings moving around in that space are characterized and identified via the ventral stream. This is binding of a very high order; it is perhaps not too far off to say that the STC is a locus for personal identity.

The STC is uniquely situated in several respects. Lying betwixt the 'how to' and 'what' pathways, it mediates both perception of objects and perception for action. It also lies at the junction of the parietal, temporal, and occipital lobes of the brain, where the multiple sensory modalities of touch, sound, sight, and smell are integrated. Finally, and most intriguingly, the STC acts as the fulcrum for two complementary sensory-motor functions: understanding one's own actions, and understanding the observed actions of others. We shall see that these two processes utilize the same machinery, but in reverse direction.

If I tell you to hold your thumb and forefinger three inches apart while I hold a dollar bill between them, and then suddenly release it (unannounced), you will probably be unable to catch it. But if *you* drop the bill with one hand, you can easily catch it with your other. Why? When we do something to ourselves, it feels different from someone else doing it to us, even if the act itself is identical. The reason is a 'feed forward' anticipation signal mediated by the cerebellum. An intention to act or move activates the cerebellum, which then attenuates subsequent sensory input from this region of the body. Thus, internally generated tactile or auditory stimuli, such as tickling, pinching, or talking to oneself, are experienced as less intense than an equivalent stimulus from someone else. This dampening of self produced stimuli is a mechanism which enables us to plan and predict the course of our ongoing actions.

Some neuroscientists have proposed that the motor loop may also work in reverse, enabling us to understand the motions of others. In other words, when we see someone moving or doing something, we create a representation of that movement in our own brains, giving us some understanding of both the sensation and the motivation behind that movement. This is called the '**direct matching hypothesis**'. It implies that understanding actions, and, by extension, understanding others' intentions, requires a direct match between observed behavior and the observer's own motor representations of that behavior. At first glance, this explanation does seem contrived, but there is mounting scientific evidence in its support.

First, neuroimaging studies have found unique activation patterns in normal human volunteers observing 'biological motion'. This includes live action movies, cartoons of people or animals, illuminated stick figures, and animated 'biomorphs'. These activation patterns are not found when the subjects view non-biological motion such as randomly moving rigid rods. The activated brain areas involved include the medial prefrontal cortex, the right STC, and the cerebellum. Autistic subjects have been found to have abnormalities in many of these brain regions.

Second, scientists working on monkeys have found certain cells in the STC that fire both when an animal is doing something **and** when it is observing other monkeys doing the same thing. These remarkable cells have been dubbed '**mirror neurons**'. They are activated by observed actions including hand grasp, finger movements, head turns, and eye movements (gaze direction). Moreover, they are somatotopically organized. A monkey watching another monkey grab a berry with its **left hand** will stimulate the mirror neurons representing its **right hand** in the STC. Through this matching process, a mirror map is generated which allows the individual to understand or simulate what she has seen. There is now some intriguing evidence that the mirror neuron system is defective in autism.

Third, there is a rare developmental disorder called **Mobius syndrome**, which is characterized by paralysis of the facial muscles, defective eye movements (strabismus), and profound social retardation, despite normal intelligence. Individuals with the syndrome have no sensory abnormality. Rather, it seems that their congenital inability to produce facial expressions and therefore represent them in their own brain's motor areas prevents them from understanding the expressions of others. Not surprisingly, Mobius syndrome is highly associated with autism.

The STC does seem to be the critical link between the perception of action and the understanding of intention. But humans have one other brain region that is also crucial for understanding intention: the **medial prefrontal cortex (medial PFC)**. Imaging studies have shown that observed biological action activates the medial PFC as well as the STC. In fact, there seems to be a circuit linking the two.

Understanding other minds through observation is a unique human ability. It allows us to judge people without direct confrontation or explicit information. It is a shortcut for dealing with complex and potentially dangerous social situations smoothly. Mentalizing, or what the psychologist Simon Baron-Cohen calls 'mindreading' distinguishes mature human beings from infants, apes, and computers. This is known as the '**Theory of Mind**' (**TOM**). It has not escaped my attention that the mirror neuron system may form the neurological basis for Theory of Mind, and much else besides.

TOM, at its basic level, is illustrated in the following cartoon of Henry and Victor (**figure 3**). Understanding this scenario requires the ability to attribute mental states to others, and forms the basis for deception and pretend play. Normal infants understand pretence and goal-directed action by age two. By age four children can, and regularly do, engage in deliberately deceptive activities like lying and cheating. Typical four year olds would be expected to pass the Henry-Victor/TOM test. Children of the same age with Asperger/higher functioning autism are likely to fail, and the more profoundly autistic will not pass at any age.

Figure 3: Henry-Victor experiment

Defective TOM is a key component in the autism spectrum disorders. It now seems to have a precise anatomic location: the medial PFC. In 1996, the neuropsychologist Francesca Happe and her colleagues at University College, London conducted a remarkable set of experiments using PET scans on TOM tasks. First they looked at brain activity in normal volunteers engaged in silently reading and answering questions on a set of three stories. One story consists of unlinked sentences. A second story describes a physical account of a burglary. The last story requires the subject to implicitly understand the mind of a character. The TOM story selectively activated the left medial PFC in normal volunteers. Next they tested Asperger subjects (otherwise matched for IQ) on the same task. The Asperger subjects demonstrated markedly diminished activity in this region.

At first glance, autism seems to involve abnormalities in two totally unrelated domains: mentalizing ability and motor control. But we can now see that the two domains are, in fact, closely related. We act on what we know or believe, and we know or believe what we see. Autism does not involve deficits in the early stages of sensory/perceptual processing, nor does it affect movement in the late stages of motor control. The imaging studies described above suggest that the PFC is not activated by its normal incoming stimuli. This could be due to defects in the STC, cerebellum, basal ganglia, thalamus, or in the connections in between. Imaging studies have found various lesions in each of these and in other brain structures, but there is as yet no consensus on autism's 'anatomic locus'. Alternatively, the core defect may be a diffuse wiring problem between these structures rather than discrete damage to any one of them.

A final question concerns the nature of clumsiness in the different varieties of autism. Some believe that autism and Asperger syndrome are separate disorders based on, among other things, their differing motor deficits. AS children have been noted to be rather awkward and gauche, while autistic children appear more agile or nimble. One theory is that the two disorders affect separate parts of the motor pathway. For instance, a primary disconnection between the motor cortex and the basal ganglia may lead to problems in the **preparation** of action, seen in the hesitant, awkward sports performance of AS children. On the other hand, a disconnection between the prefrontal cortex and the basal ganglia might produce problems with **choosing** an action. This is characteristic of more classically autistic children.

I would like to propose an alternative hypothesis: that classic autism and AS are mainly cortical disorders that differ in the quantity of cortical dysfunction. Classic autism is generally more severe in terms of cognitive function because more cortex is damaged, but less severe in terms of motor function because the basal ganglia take over some of the movement normally mediated cortically. The cortex is like a chief executive making the decisions, taking on novel or difficult jobs, and generally delegating authority to semiautonomous underlings for routine day-to-day operations. The basal ganglia are recruited for well-learned or 'automatic' actions, such as riding a bicycle. If the cortex is too dysfunctional to control volitional movements, then perhaps the basal ganglia take over, producing the smoother, but more stereotypical and repetitive behavior seen in autism. In contrast, AS individuals retain enough cortical function to control voluntary movements, but are not very good at it, resulting in hesitant, clumsy movement.

◆   ◆   ◆

'Are you a medical student?'

'Uh, yes,' I say wearily. I have assisted on three cases already, been yelled at twice, and nearly fallen asleep holding a clamp over a cirrhotic liver.

I look up from tying my sneaker laces in the OR locker room. It's Dr. Kauffman, chief of neurosurgery.

'How'd you like to assist me on a case?'

I look down at my watch. It is almost 10 PM. I should be at home studying for my surgery clerkship exams next week.

'When?' I ask, trying to sound eager.

'Five minutes.'

'Uh, okay.'

'Good. It's great to have eager students like you! It'll be an easy case. Intracranial bolt. The guy's bleeding into his brain. Let's scrub in.'

Another fine mess I got myself into. Okay. Just calm down, relax, do what you're told, he won't expect too much from you. Hell, he might even give you a recommendation.

Kauffman enters the OR and puts up some MRI scans on the viewbox.

'Alright, ladies and gents, this is Dr. Kong. He will assist us on this case.'

I sense everyone's eyes turn menacingly on me.

'Mr. Dagastino here, a 68 year old obese hypertensive smoker non-compliant with his meds had a massive intracranial hemorrhage into his left basal ganglia two days ago. You can see blood compressing this structure here. What is that structure, Dr Kong?' he asks without looking at me.

'The lateral ventricle?' I reply sheepishly.

'The lateral ventricle. The lateral ventricle. Is that correct, Miss Harrington?' he says looking over at one of the scrub nurses.

'No, I think that's the third ventricle.'

'The third ventricle and the Aqueduct of Sylvius. That's absolutely correct Miss Harrington. Dr. Kong, I suggest you study your anatomy a bit harder.'

I feel like a fucking moron.

'Our patient is starting to emerge from his coma thanks to the steroids and mannitol, but his MRIs show an increase in the bleed. We may need to put in a drain for decompression. All you have to do is hold this intracranial bolt until I drill the burr hole. Then we screw it in place, hook it up to the wires, and check out his intracranial pressure. Very simple. Okay?'

'No problem,' I say.

Kauffman finishes drilling through the skull in a few minutes. I give him the little plastic bolt. It doesn't quite fit yet. While he's fiddling with it, my mind starts to wander. I'm thinking about the third ventricle and how it looks different on the MRI compared to my neuroanatomy text.

'Take it. Take it! TAKE IT!' Kauffman yells at me handing over the bolt.

I move my right hand towards him, but at an awkward angle. The bolt slips through my gloved fingers and falls to the floor with a bounce.

'Goddammit!!'

The room becomes very quiet.

Kauffman lets out a long sigh and turns to Miss Harrington.

'Do we have another one here?' he asks calmly.

'No,' she replies in a soft Irish accent.

'Allright, then call CDPD and ask for another. We'll just have to wait.'

I feel mortified. For a second, I am overcome by vertigo and almost fall.

Kauffman looks at me with squinting eyes but doesn't say anything. The rest of his face covered by his surgical mask.

Miss Harrington seems to be staring at me with a hateful look as well.

My hands are trembling. After what seems like an eternity, the second bolt comes up. This time, the neurosurgeon entrusts it to Miss Harrington.

◆   ◆   ◆

Summer 2001. On a glorious morning, I ride my bike north from Battery Park. Just before the anchorage of the New York Yacht Club, with the impossibly gigantic twin towers of the World Trade Center to my right, is a volleyball court. A group of young people is in the middle of a pick-up game. Most of the men are shirtless, revealing rippling abdominal muscles. The women are wearing colorful skintight lycra shorts and support bras over well toned bodies. They're all in great shape. For a second, I feel a mixture of envy and self-disgust. But it fades. In a moment, I am speeding past them with the wind on my back. I'm going to have my own fun today.

After a long bike ride up to Morningside Heights, I feel great. So what if I can't play volleyball? My life is for me to live. That afternoon, at the Columbia University library, I come across an interesting paper in a journal devoted to autism research. The authors had devised a self-administered test (which they call the 'autism spectrum quotient' or 'AQ') designed to find where along the 'autistic spectrum' a given individual might fall. [**Baron-Cohen, Simon, et al. 'The Autism-Spectrum Quotient (AQ): Evidence from Asperger Syndrome/High-Functioning Autism, Males & Females, Scientists & Mathematicians' in <u>Journal of Autism and Developmental Disorders</u>, Vol. 31, No. 1, 2001**]

The test consists of 50 questions designed to probe different aspects of the autistic phenotype, such as imagination, socialization, communication, executive control, theory of mind, and central coherence. It is scored from 0 (no autistic characteristics) to 50 (extremely autistic). The average score is 16. Males, scientists, and mathematicians tend to score slightly higher than females, artists, and social scientists. People with autism/Asperger syndrome average about 36. I eagerly take the test. My AQ comes out at 38.

4

Anxiety

Lehman Hall, Harvard Yard, late October 1987: 'I feel fine and I feel good. I'm feeling like I never should…'

Strobe lights spin out psychedelic shadows on the wall through smoke-filled air like stained glass windows in a crazy cathedral. The DJ is playing New Order's 'Bizarre Love Triangle', for some reason, a perennial favourite with the Asian crowd. It is 10 o'clock and the dance floor is still half-empty. Harvard isn't known for its undergraduate parties, but losers can't be choosers. Three young women on the other side of the room stand up and start dancing. I'm standing in the dark, nervously downing a beer. Waves churn in my stomach and my palms start to perspire. The song goes on, '…whenever I get this way, I just don't know what to say. Why can't we be ourselves like we were yesterday…'

I hesitate on the brink of my next step. The feeling reminds me of entering the cafeteria on the first day of first grade; I was sick to my stomach from the frightful noise and the hopelessly high ceilings. I cried so hard that Mrs. Dixon had to take me into the principal's office. I had it again at my college freshman orientation picnic. I had to force myself to look calm attempting to make sandwiches while introducing myself to complete strangers. I can recall the sensation of gagging on dry roast beef when I tried to talk with my mouth full. This sort of thing happens to me all the time.

I've been secretly going to off-campus parties lately. The anonymity gives me a sense of security. Maybe I can pretend to be the cool and popular guy I'm not. But it rarely works out that way. I usually start out well enough with some pre-rehearsed lines and a ready smile. But I soon run out of things to say. The rest of the conversation is excruciating as I try to save myself from embarrassment. Seeing pretty young girls smiling and chatting away with boys fills me with immense frustration, jealousy, and anger. What on earth could they talking about? I am just six inches away from their faces, but they might as well be a million miles away. How do I connect? Why is it so hard to act normal?

I step across the threshold of the dance floor and try to sway to the rhythm. But my awkwardness must be obvious, because the girls look at me and laugh. I approach the slim blond one in the tight black skirt and shout in her ear, "IS IT OK IF I DANCE WITH YOU GUYS?" Just then, I can feel my heavy eyeglasses starting to slide down my sweaty nose. I quickly push them up with my index finger and wipe my forehead. She says nothing; she doesn't even look at me, but whispers something to her friend. They start laughing, and then, inexplicably, leave the dance floor in mid-song. That leaves me standing alone, confused and embarrassed, still awkwardly shaking to the New Order: '…I'm not sure what this could mean, I don't think you're what you seem…' Suddenly I realize there's a group of people staring at me. I recognize them from MIT. One of them is my roommate. What the hell are they doing here? Mortified, I weave my way through the dancing bodies towards the exit, pretending not to have noticed them. I feel a burst of relief as I run out into Harvard Square and the cold autumn air chills my sweat soaked hair. "Goddamit! Fuck! Fuck! Goddamit! Fuck! Fuck! Goddamit! Fuck! Fuck! Goddamit! Fuck! Fuck!…" I mutter loudly and repeatedly to myself as I run past the young punks with spiked mohawks and eyebrow rings at the top of the Harvard T stop escalator. They pay me no attention.

◆   ◆   ◆

Looking back on it now, those years that should have been among the best of my life seem full of stress and anxiety. So much effort gone into trying to feel and act 'normal'. The most stressful thing was not the biochemistry I was supposed to be learning, but the social chemistry that I wasn't. While others could be popular, have girlfriends, and really enjoy themselves, I was always trying to figure out what I could do to be more like them. I had a lot of problems, but I now realize that the biggest one was not at all obvious. There were plenty of other short and chubby students with thick glasses, poor athletic skills, and a geeky affinity for the library, yet many of them did quite well in the social department. After all, this was MIT. My problem was anxious withdrawal.

The anxiety arose from two sources: fear of the unknown and a self-fulfilling expectation of failure. Others' thoughts and behaviors are often mysterious and therefore threatening. Is there a hint of disdain or hostility lurking behind those eyes and that slightly crooked smile? Am I paranoid in thinking there is something dismissive in her tone of voice? The expectation of failure comes from a history of failure (real or imagined) and of criticism (whether deserved or not). It is something that lingers on in the subconscious like a mildly unpleasant odor in the kitchen. It does occasionally disappear when I feel that I'm in complete control of the situation. But because of the dual nature of this anxiety, I usually did not reach out to make friends. It seemed that the better I got to know someone, the more anxious they would eventually make me feel.

Paradoxically, I have always felt most comfortable around strangers. Some of my most enjoyable experiences have been backpacking alone through strange and wondrous European cities, not knowing the people, the languages, or the customs, with only a street map to guide my steps. Yet more than anything those were the experiences of my youth that I remember as times of 'completeness'. These were times when I would imbibe the ebb and flow of culture and humanity without self conscious anxiety. I may not have known the people I met, but they didn't know me, either. So we were even. The problem started once they got to know me; then the familiar pattern would reassert itself. I would have to leave to regain my self confidence.

Emotions and Feelings

There are **emotions**, and then there are **feelings**. In casual speech, we often use the two terms interchangeably. But for our (more formal) purposes they refer to two quite different things. Emotions are physical signals, elemental and visceral. Fear, anger, happiness, grief, surprise, anxiety and various combinations thereof comprise our emotional repertoire. In essence, emotions refer to **the physical state of arousal** produced by a particular stimulus. They are simply the body states caused by the action of the nervous system and parts of the endocrine system, which together oversee many of the unconscious regulatory activities necessary for life. For example, the rapid heartbeat and breathing, clammy hands, and dry mouth some of us experience before taking an exam constitute the emotion of anxiety. Being conscious of this emotion makes it a 'feeling'. A feeling is the **conscious reflection** of an underlying emotion. Emotions by themselves are not conscious. But both unconscious emotions and conscious feelings guide our behavior.

Feelings are complex elaborations of underlying emotions. They involve a mixture of emotion and reason. This happens when we feel (or say we feel) guilt, pride, embarrassment, or schadenfreude. Contrary to common belief and intuition, feelings are usually the result of emotions, rather than the other way around. As the great American psychologist William James put it, 'We feel sorry because we cry, angry because we strike, afraid because we tremble and not that we cry, strike or tremble because we are sorry, angry or fearful as the case may be.' Of course, thinking sad thoughts can make us cry, and dwelling on past injustices can make us angry, but often it is the subconscious emotional state that brings these thoughts to the surface.

Emotions evolved in our animal ancestors because they are useful for survival. Organisms use them to represent and communicate information about the most basic drives of life: food, sex, and health. In humans, they have developed into something more elaborate: feeling states, which allow better perception and control of social behavior. Emotions are the domain of a part of the brain called the limbic system, which evolved with the first mammals over 200 million years ago. All mammals have limbic structures in their brains, presumably to facilitate the communication demands of a warm blooded, fast-paced lifestyle. Emotional expression allows social animals to reveal their needs and desires to themselves and to others in a more 'compact' and abstract form. Rats, cats, and bats all display anger, fear, surprise, and presumably joy through facial expressions designed to be readily visible to others. Organisms capable of these expressions can convey much more information about their physical and biological state than organisms that are not. Reptiles, fish, and other cold-blooded animals rely largely on more direct, less 'symbolic' signals to achieve these ends. They eat and sleep, and often see and hear exceptionally well. But fish eyes and reptile eyes are cold. Unlike mammalian eyes, they are devoid of happiness, sorrow, fear, or anger. We will briefly examine the neuroanatomy and physiology of the primate limbic system before moving on to the elaboration of feelings in human beings.

The key component in emotional processing is the amygdala, a small ovoid structure deep within the medial temporal lobe on each side of the brain (**figure 1**). The amygdala receives 'raw' sensory input (visual, auditory, somatosensory) either directly through the brain's great relay station, the thalamus, or 'preprocessed' via the cerebral cortex. If the input is emotionally significant or 'salient', such as the sight of a snake (which seems to trigger a universally preprogrammed circuit in newborn brains), or the roar of a fire engine, the amygdala is aroused and sends an activating signal to the hypothalamus. The hypothalamus, in turn, activates the sympathetic system (part of the autonomic or 'automatic' nervous system), which increases heart rate, blood pressure, and reflexes, and dilates the pupils, preparing the individual for action. This is the so-called 'fight or flight response'. At higher levels of negative emotional arousal (fear), the parasympathetic system (another part of the autonomic nervous system) is also activated, producing contractions of smooth muscles leading to the rather unpleasant loss of bowel and bladder control. With sustained anxiety, the hypothalamus instructs the endocrine system to secrete 'stress hormones', the corticosteroids, which have multiple, far ranging actions on the body, including immune system down-regulation, changes in salt and potassium balance, and psychological functioning. In effect, the long-term health of the body is sacrificed for enhanced short-term performance. (**figure 2**)

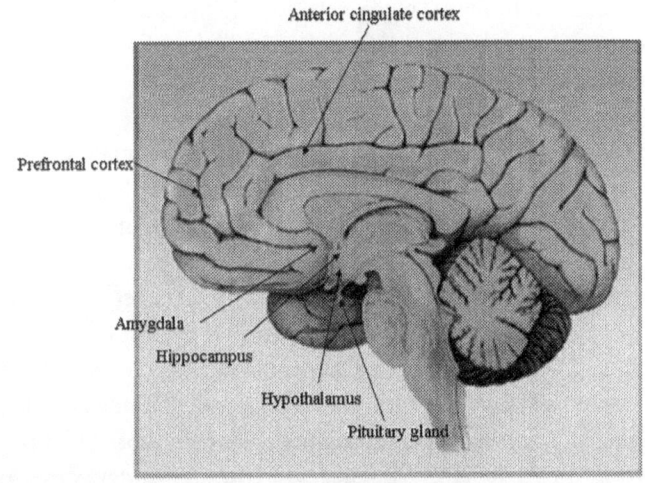

Figure 1: amygdala

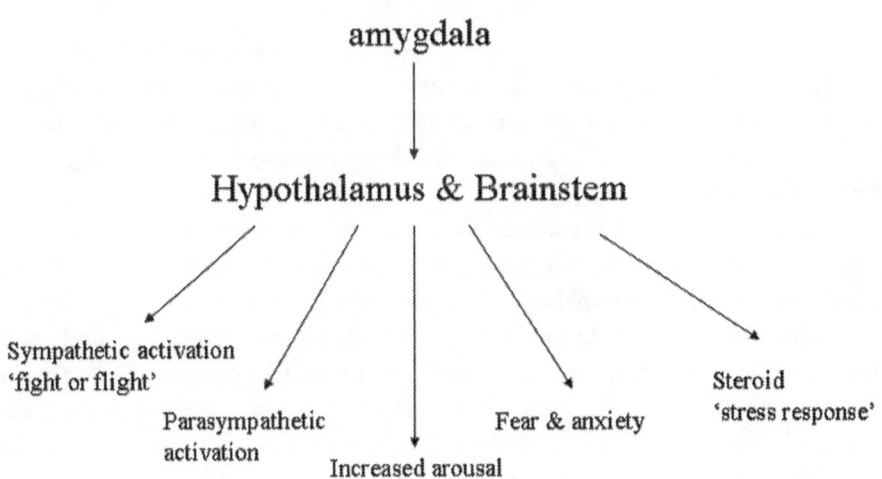

Figure 2

In this way, the amygdala generates an 'emotional valance' tagged onto particular stimuli or situations that the organism has learned to associate with something good or something bad. Once the initial learning or conditioning phase is completed, and the emotion is 'bound' to the corresponding perception of the object or situation, the emotional trace or memory is stored in the amygdala or surrounding cerebral cortex. Whenever the creature subsequently finds itself in a similar situation, it automatically and unconsciously generates the appropriate emotional response. In humans, many emotions (generated by the amygdala) are transferred up to the cortex where they enter the realm of awareness and become **feelings** to be utilized for social judgment and behavior.

The major input to the amygdala is sensory (of all modalities: sight, sound, touch, taste, smell). The output from the amygdala is twofold. First there is the 'descending' channel to the hypothalamus and brainstem responsible for the emotional response (fight or flight). Second, there is an 'ascending' channel to the limbic cortex, which produces feelings. The limbic cortex refers to several regions in the front or 'prefrontal' part of the brain, which receive massive connections from the amygdala. The **prefrontal cortex** is the largest and most highly evolved part of the primate brain. It is responsible for the calculations and decisions that facilitate rational thought and social judgment, namely those attributes we commonly refer to as 'intelligence'. We will revisit the prefrontal cortex in terms of nonsocial cognition in chapter 5. The amygdala is connected to a particular part of the prefrontal cortex: the **medial prefrontal cortex** (abbreviated **mPFC**).

First, the images of direct sensory perception as well as internally generated 'images' (imagination) arrive in the amygdala and are passed along to the 'body' via the hypothalamus and lower parts of the nervous system. This is what the neuroscientist Antonio Damasio in his thought-provoking book, **<u>Descartes' Error</u>** calls the 'body-loop'.

Second, output from the amygdala modulates the mPFC and produces conscious feelings that accompany the emotions produced by the body-loop. This is what I like to call the '**brain-loop**' (Damasio calls it the 'as-if body loop') because the feelings in the brain do not require the body to experience the emotions to make them real. In other words, a stimulus or image with emotional valance can be processed directly by the cortex without having to first pass through the body. We can be conceptually afraid of heights, even if we are not standing at the edge of a cliff. We can still love our spouse, even if the sight of her does not quicken the pulse.

Even though feelings (the brain-loop) do not require emotions (the body-loop) to be apprehended, they would pale without them. Emotions simply heighten whatever it is one is feeling, or thinking. Subconscious emotions are a powerful influence on what and how we feel. And somewhat paradoxically, emotions and feelings are essential in helping us make decisions. The body loop produces what Damasio calls '**somatic markers**': the emotions associated with each choice that help us decide between them. The body-loop and the brain-loop join in the mPFC producing the feelings necessary to expedite the decision-making process.

There is an optimal level of emotion that is necessary to make good choices. Damasio describes a patient, Elliot, who unfortunately suffered from a brain tumor that compressed parts of his prefrontal cortex and limbic system. The tumor rendered him incapable of feeling emotions. Additionally, Elliot became totally indecisive. Despite an intact intelligence and insight, he was unable to manage his time and prioritize sets of tasks. Asked to go buy some items at the local grocery, he would spend hours deliberating which brand of cereal to get.

I have the opposite problem. My emotion tends to run amok. It doesn't take much to make me jump with delight or fume with anger. I tend to want to purchase the latest gadget I read about, or believe the latest gossip. My emotions chain my decisions and choices to whatever is close at hand. Both Elliot and I make poor decisions, but for opposite reasons. Emotions prime us to choose among the many possible outcomes represented in our cerebral cortex. Too little emotion creates indecision, too much emotion leads to impulsivity.

The mPFC lies on the front inner surface of the brain between the two cerebral hemispheres. It has many connections to other brain regions, including the thalamus, basal ganglia, amygdala, and hippocampus. But perhaps the most important connections involve ascending fibers from nerve cells deep in the brainstem. Some of these cells secrete **dopamine**, the neurotransmitter that plays an important role in the modulation of movement by the basal ganglia. Dopamine and its close chemical cousins, **serotonin** and **norepinephrine**, also secreted in the brainstem, are called **neuromodulators** because they influence the behavior of large sets of neurons all over the brain. For instance, after the successful completion of certain appetitive activities such as eating or sex, serotonin released from the brainstem neurons floods the cortex and produces a relaxed, sleepy, content state. This is why drugs that interfere with removal of serotonin from the synapses (thereby increasing the active serotonin concentration) known as selective serotonin reuptake inhibitors, or SSRIs, are such popular antidepressants. In the prefrontal cortex, dopamine activates cortical neurons, enhancing the thought process and producing a feeling of 'reward'. For this reason, the dopamine system is said to confer 'value' to thoughts.

Bear with me, because this is where it gets interesting. The dopamine value system **activates** the prefrontal cortex, but tends to **inhibit** the amygdala. In fact, cortical activity in general suppresses the amygdala. This is because a major neurotransmitter receptor in the amygdala, the GABA receptor, is inhibitory. When cells from the cortex or brain stem fire onto neurons in the amygdala, they activate the GABA receptors which then actually decrease the usual amount of activity in the amygdala. Thus input to the amygdala causes inhibition, while output from the amygdala turns on other brain regions. On the other hand, when there is a lack of incoming activity, the amygdala tends to become more active. Thoughts inhibit emotions, but emotions promote thought.

◆   ◆   ◆

We can now start to build a model of the amygdala-mPFC complex (**figure 3**). In this model, the amygdala is stimulated by perceptual input. It then generates emotion via the body-loop and feelings via the brain-loop. These loops converge in the mPFC and provide the somatic markers for the decision making process carried out elsewhere in the prefrontal cortex. Rewarding choices and goal-directed actions increase dopamine levels, which initially activate the the PFC. Activity in the PFC then inhibits the amygdala. This is the concept of '**feedback inhibition**'. The more active the mPFC, the less active the amygdala; the less active the mPFC, the more active the amygdala. The amygdala and the cortex are intimately linked in an elaborate feedback dance.

Normally, any stimulus associated with a dopamine surge initially stimulates the amygdala, promptly resulting in an emotional response. This, in turn, produces a conscious feeling of joy or shame, excitement or fear, depending on the associated memories and instincts brought out by the emotion. The feeling then sublimates the earlier emotion, dampening its initial effect.

I am less affected by this feedback inhibition. I am quick to anger, quick to forgive, quick to celebrate, and quick to forget. Any minor insult or critique (whether real or imagined) produces a powerful visceral reaction that quickly seems to overwhelm and shut down the 'thinking parts' of my brain. I simply become more excited, more frustrated, or more distraught without really knowing why. My emotions remain raw and volatile, without the feelings to justify them.

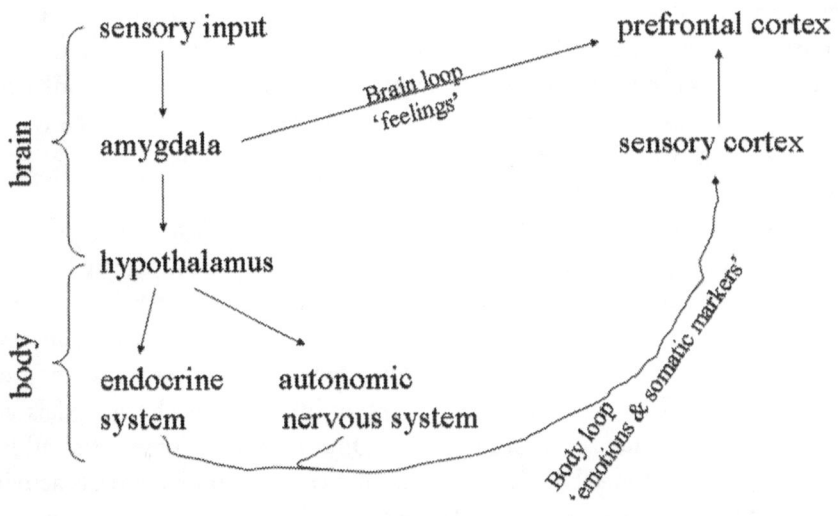

Figure 3

Emotional learning

The amygdala is a way station between the temporal lobe on the one hand and the hypothalamus and mPFC on the other. This remarkable little subcortical lump of nervous tissue straddles the spheres of perception, emotion, and cognition. Damage to the amygdala disables the ability to perceive the significance of emotional displays in others and dampens one's own emotional output. For instance, when subjects with bilateral amygdala damage are shown pictures of auto wrecks, wartime gore, and the like, they exhibit no increased heart rate or sweaty palms (as measured by galvanic skin response, the basis of lie detector tests). Normal people do.

Jocelyne Bachevalier at the University of Texas has studied the behavior of baby monkeys after the experimental removal of their amygdalas. These animals became more passive and socially withdrawn as they matured. They failed to recognize and were indifferent to signs of aggression or friendship from other monkeys. In addition, the animals were found to display little eye contact or spontaneous body expression, but did display increased stereotypic motor patterns such as rocking and hand flapping. In short, these observations seem to provide an animal model of human autism. Can a dysfunctional amygdala alone explain the autistic disorders? The evidence is mixed and far from conclusive. While pathological changes of the limbic system, including the medial temporal lobe and amygdala, have been described in autistic individuals, it is unclear that those changes are actually responsible for the constellation of characteristic abnormalities. A more likely explanation is that the core deficit of autism lies somewhere between the amygdala/medial temporal cortex and the prefrontal cortex. This idea becomes more compelling when we examine the interconnections of these regions with respect to emotional learning and anxiety.

The amygdala is the site of emotional learning. It is where visual and other incoming sensory information converges with ascending value signals from the brainstem (dopamine secreting neurons) to bind features together into an emotionally coherent whole. The amygdala paints the observed world with an emotional hue. The process of emotional learning is most important in infancy and early childhood. This is when social perception mediated by the amygdala and medial temporal cortex is linked to social cognition further along in the mPFC. The feelings that aid social decision making are the products of amygdala activity. Without the amygdala to confer the 'emotional tag' to a familiar face, for instance, the proper development of social reasoning is arrested.

Autism and Emotional Perception

The temporal lobe is closely connected to the limbic system structures of the amygdala and the mPFC. In addition, it is the site for higher-order visual processing, especially of faces. It is well established that part of the temporal cortex called the **fusiform gyrus** (FG) is activated by the sight of faces. (**figure 4**) Individuals with the unusual misfortune of having selective damage to this area on both sides as a result of strokes, tumors, or congenital abnormality, often show an isolated inability to recognize faces, even of family members. This curious disability, called prosopagnosia, can exist with otherwise intact cognitive, perceptual, and emotional functions such as recognizing and naming buildings or animals, reading, remembering names, or demonstrating affection for loved ones (once pointed out). Moreover, researchers have found that certain cells in the FG of monkeys are selectively activated when shown faces or face-like visual stimuli. These so-called 'face cells' form part of a social recognition module at the high end of the visual processing stream, and are connected to the limbic system.

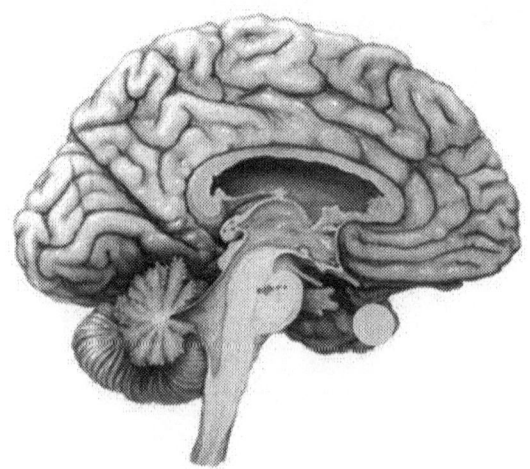

Figure 4: fusiform gyrus

Neuroimaging studies have been done comparing normal subjects to autistic and Asperger syndrome individuals on face observation tasks. They have revealed that the FG is specifically activated in the control subjects, while the autistic/AS subjects displayed much less activity in the FG, but demonstrated greater activation in the surrounding areas of the temporal cortex. These surrounding areas were equally activated in all subjects during viewing of nonface objects such as tables and automobiles. This seems to indicate that autistic subjects perceive faces using parts of the brain that normally process objects. Unlike normal subjects, autistics may not have a fully developed specialized neural system dedicated to tagging and processing face-like stimuli.

These findings tie in with the observation that autistic people seem to have better object perception than social perception, or that they concentrate on 'things', rather than on people. Perhaps a causal factor for autism lies in the fusiform gyrus. But the deficits in autism go far beyond simple social misperception. They also affect emotional expression and thought.

Social cognition, the ability to make sound social decisions throughout the lifespan, also requires the activity of the mPFC to which the amygdala projects. There is a neural fiber circuit that runs from the temporal lobe through the amygdala to the mPFC. In autism, part of this circuit may be cut off, perhaps somewhere between the amygdala and the mPFC. This roadblock prevents the normal activation of the cortex by the amygdala, while simultaneously **preventing the normal inhibition** of the amygdala by the cortex. (**figure 5a&b**)

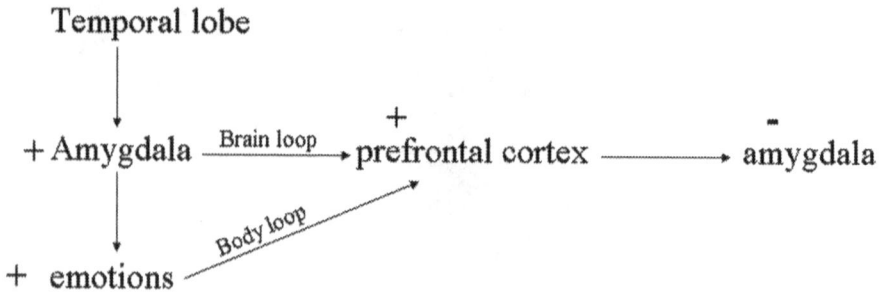

normal

Figure 5a

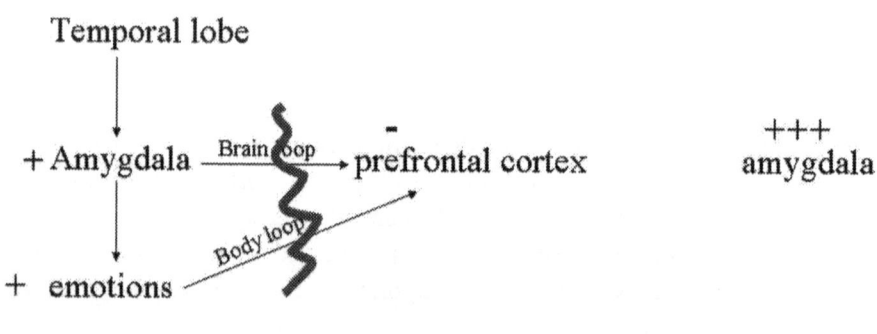

autism

Figure 5b

Anxiety

Recall the process of frontal inhibition: the cortex tunes down the amygdala, while the amygdala turns up the cortex. If you will forgive a very simplistic analogy, 'reason' suppresses 'the passions' while passions stimulate reason. If the cortex is no longer functioning, there is no 'off' signal to the amygdala, and it will be constantly turned on. The balance between the mPFC and the amygdala (the cortical and subcortical components of the limbic system) is lost. The mPFC, the mediator of social cognition and the interpreter of emotions, becomes less active while the amygdala, the engine of emotions, especially fear and anxiety, is hyperactive. This may play a role not just in the autistic spectrum disorders, but also in other conditions such as attention-deficit disorder, major depression, generalized anxiety, and perhaps even Parkinson's disease.

In a series of experiments demonstrating frontal inhibition, professor Joseph LeDoux of NYU found that monkeys with lesions to the mPFC could be taught conditioned fear (associating an auditory tone with an electric shock). This is to be expected as the animals have fully functioning amygdalas which allow normal emotional learning. However, when the conditioned stimulus (the tone) was regularly presented without the noxious stimulus (the electric shock), the animals failed to extinguish the fear response. In other words, the animals had become hyper-anxious. The defective PFC was unable to turn off the amygdala.

Anxiety is a negative emotion, the result of an overactive amygdala. It is also associated with reduced activity in the cortex and low dopamine and serotonin levels. If anxiety is a product of an underactive PFC and an overactive amygdala, can drugs targeting this imbalance be used to treat anxious states? It turns out that they can. The first arm is to increase dopamine levels to activate the PFC. Stimulants such as amphetamines and ritalin do this by releasing dopamine stores in the PFC. Ritalin is successful in treating attention deficit disorder (ADHD) and certain types of depression. Its efficacy in the treatment of autism remains unproven.

A major problem with the stimulants is their potential for addiction and abuse. In fact, many drugs of abuse from nicotine to caffeine to cocaine work by flooding the brain with the reward chemical, dopamine. The reason that it feels so good (for some) to smoke a cigarette, sip a cup of tea, or snort a vial of crack is fundamentally the same reason that it feels so good to pass an exam, climb a mountain, or have good sex. On the other hand, the SSRIs such as paxil, celexa, and prozac, which increase the concentration of dopamine's sister compound, serotonin, are quite effective at targeting the anxiety associated with both autism and depression without some of the dangers of addiction associated with dopamine surges.

The second arm in attacking anxiety is to suppress the amygdala. As we have seen, a major player in the amygdala is the GABA receptor. Stimulating these receptors will turn down amygdala output and quell anxiety. Activating the PFC with dopamine or amphetamines will do this indirectly, but so will direct stimulation of the GABA receptors with drugs called benzodiazapines. For this reason, the 'benzos' such as valium and atavan are among the most used (and abused) medications. Like the amphetamines, the benzodiazapines have high addictive potential, and do not have a solid role in the pharmacological treatment of autism.

The Theory of Mind

We have now come to the high point of emotional processing, the mPFC. We've glimpsed its landscape before: it is the site of the Theory of Mind Module (TOMM), the part of the human brain that represents one's beliefs and compares them to information about others' beliefs, to synthesize 'beliefs about beliefs' or 'representations of representations'. TOM refers to understanding that other minds have separate belief systems apart from one's own. Arriving at a TOM requires that one comes to understand others' intentions and beliefs. Mentalizing depends on picking up emotional cues such as hand gestures, posture shifts, gaze direction, and facial expressions and linking them to the otherwise invisible contents of another mind. This is the essence of emotional perception, and it is what is missing at the core of autism.

The perception of faces mediated by the fusiform gyrus can be thought of as representations of 'the other'—that which is not oneself. These are then compared to the representation of the 'self' in the mPFC. The contents of the 'self' come largely from the amygdala directly through the 'feeling' of the emotions it generates. They can also come indirectly, through the somatosensory perception of one's bodily sensations such as palpitations, perspiration, a 'tingle up the spine', or 'butterflies in the stomach'. These, in turn, are the products of the autonomic nervous system and endocrine system connected to the hypothalamus and amygdala. The direct path to the self is the brain-loop, the indirect path is the body-loop. The two loops originate in the amygdala/medial temporal lobe and end up in the mPFC (**see figure 3**).

The continuously updated (subconscious) comparison of one's 'self representation' with 'non-self representations' in the neural networks of the mPFC forms the basis for understanding others' intentions (TOMM). A working TOMM, in turn, allows one to manipulate that understanding for pleasure and for profit (pretense, deception) and to give comfort to others (empathy). These remarkable abilities are routinely mastered by the vast majority of normal four and five year old human children all around the world.

There is more. Better understanding others enables us to differentiate ourselves as discrete unified agents anchored in space and time capable of exercising focused attention and freely willing goal-directed actions. As the adage goes, 'to know others is to know ourselves' and vice versa. Knowing oneself gives one the focus, the reference, the object, and the motivation for one's actions. It provides 'will power'. But where there is a will, there must be a way. There is. That 'way' is also to be found in the PFC. The process of decision-making is the preserve of the nearby **dorsolateral prefrontal cortex (DLPFC)**, which we will examine in more detail in chapter 5.

Defects in Emotional Communication

Understanding others, understanding oneself, and acting on behalf of one or the other are jobs for the mPFC. The perception and expression of emotions are a type of information transfer, the language of the limbic system, if you will. Communicating in any language involves three basic steps: listening, understanding, and speaking. So it is with the language of emotions. One must first listen (perceive the emotional input from others via the temporal cortex and amygdala), understand (integrate the data into a coherent whole in the mPFC), and finally talk (express thoughts and desires through emotional display). Defects can be expected to occur at multiple possible points in this pathway. We will compare autism/AS with a fascinating disorder called Williams Syndrome to get a better idea of the dissociable nature of emotional perception and emotional understanding.

Williams syndrome is a rare genetic disorder associated with a deletion of several genes in chromosome 7. 'Williams people' as they are affectionately called, are characterized by distinctive facial features which make them look a bit like elves. They suffer a high incidence of congenital heart abnormalities, are usually mentally retarded, and most remarkably, demonstrate very precocious and vividly expressive language. The content and quantity of language in Williams people belies their usually low IQ. Professor Ursula Bellugi, director of the Laboratory for Cognitive Neurosciences at the Salk Institute has studied language development in Williams Syndrome. Here is an example of a 17 year old with Williams Syndrome, IQ 50, describing a cartoon:

'Once upon a time when it was dark at night, the boy had a frog. The boy was looking at the frog, sitting on the chair, on the table, and the dog was looking through, looking up to the frog in a jar. That night he sleeped and slept for a long time, the dog did. But the frog was not gonna go to sleep. The frog went out from the jar. And when the frog went out, the boy and the dog were still sleeping. Next morning it was beautiful in the morning. It was bright, and the sun was nice and warm. Then suddenly when he opened his eyes, he looked at the jar and then suddenly the frog was not there. The jar was empty. There was no frog to be found.'

And the same cartoon as described by an 18-year-old individual with Down's Syndrome with an IQ of 55:

'The frog in the jar. The jar is on the floor. The jar on the floor. That's it. The stool is broke. The clothes is laying there.' [**Lenhoff, H., et al, Williams Syndrome and the Brain, Scientific American, Dec 1997, 68-73**]

For all their eloquence, however, the content of Williams speech is superficial and devoid of significance. They do not talk to convey ideas and feelings, but rather, it seems, to pass the time much as one would mindlessly doodle while on the telephone. Much of their speech is not goal directed, and they often make things up. This largely explains why Williams people are usually unable to form lasting friendships or make organized decisions for themselves. However, Williams children are described as warm and outgoing. They enjoy human contact and, in fact, seem to display a great deal of empathy towards others. In this sense, they are like the polar opposites of AS individuals, who are socially inept, aloof, and/or anxious.

The psychologist Helen Tager-Flusberg has examined Williams children on standardized tests of TOM as compared to Down's syndrome children. She found that the Williams children performed just as badly on the interpretation of others' intentions. How can their otherwise relatively normal social expression be reconciled with their failure to understand TOM? She proposes that the perception and the understanding of social behavior, which together comprise the common notion of TOM, are actually related, but distinct, psychological processes.

Developmental disorders such as Williams syndrome, Asperger syndrome, and autism produce psychological effects through anatomical and physiological defects in different parts of the emotional circuit. Williams syndrome can be considered to involve problems in social understanding with relatively intact emotional perception. Thus, it may involve pathology in the mPFC. AS is characterized by a reverse set of problems, namely defective emotional perception with better emotional understanding. Perhaps AS involves selective pathology in the medial temporal lobe or amygdala. Finally, autistic people have defects in both domains. The problem may involve more diffuse damage throughout the limbic system. (**figure 6**)

	Williams syndrome	Asperger syndrome	autism
Emotional perception	+	-	-
Social understanding	-	+	-

Figure 6

This is, however, clearly an oversimplification. While there is modularity in the brain, the modules interact in very complex and unpredictable ways. They are not like the little balls connected with sticks you find in chemistry sets and tinker toys. Any psychological assessment of real-life disorders must acknowledge this complexity. Syndromes such as Williams and Asperger have fuzzy borders. Any attempt to deliberately pigeonhole them into a convenient theory or model runs the risk of misunderstanding their true nature. Individuals with AS generally tend to pass TOM tests in laboratory settings, but they often have profound problems with social understanding in real life. The variability of deficits across individuals on the autistic spectrum reflects, among other things, the patchy nature of the disorders themselves, and variable expression in different personality types.

Not all social interaction relies on emotional perception. Much of human behavior can be described as mechanical or behavioral stories. Most of us can choose to be as much 'folk physicists' or 'folk behaviorists' as 'folk cognitive psychologists'. But it is exclusively at the level of physics and behaviorism that autistics live. They can communicate and, in the case of the higher-functioning or AS individuals, are sometimes quite articulate. They can recognize faces and label people as friends, lovers, or enemies based on their appearance and even their expressions. But without the capacity to link emotional perception to mental states, their view of others remains very limited. As Uta Frith puts it in her book, **Autism: Explaining the Enigma**:

'...by taking account of others' mental states in even the most trivial social encounters, we exhibit what one might call affective contact. People constantly guess and monitor what each other's thoughts might be. This does not involve hard thinking, but, on the contrary, renders it easier to make sense out of a large amount of disparate information at once...

'...the failure of ordinary intentional communication and the inability to relate affectively to others are really one and the same thing. The failure is most strikingly demonstrated by the fact that even in very able autistic people understanding of language as well as of social and affective relationships remains literal. Since remarks and interactions are not experienced as part of a web of implicit presuppositions, interpretation of verbal and non-verbal signals is extremely restricted.'

Emotional perception is an effective way to come up with mindreading ability. However, not all TOM depends on emotional input. One can understand others' intentions without necessarily getting the gist of their emotional display. On the other hand, emotional perception does not necessarily imply the existence of TOM. After years of observation and practice, I can now understand when others are upset, amused, or perplexed by their tone of voice, body movements, or the like. But it's still not so easy for me to use this to predict their behavior. Unless I'm explicitly told what the other person is feeling or planning to do, they might as well be wearing a poker face. Mindreading is certainly not my forte.

Autistic and AS individuals do have access to rational thought and often use it to arrive at social judgments. Sometimes they are successful, sometimes not. Autistic people have been found to engage in inappropriate social behavior. They are chronically teased and bullied at school. They are too frequently the naïve victims of bad business deals. This is because social judgments are best made with the help of emotional modulation. These judgments can also be made through an indirect path that bypasses the limbic system, but this route, taken by those with autism/AS, is much less effective. (**figure 7**)

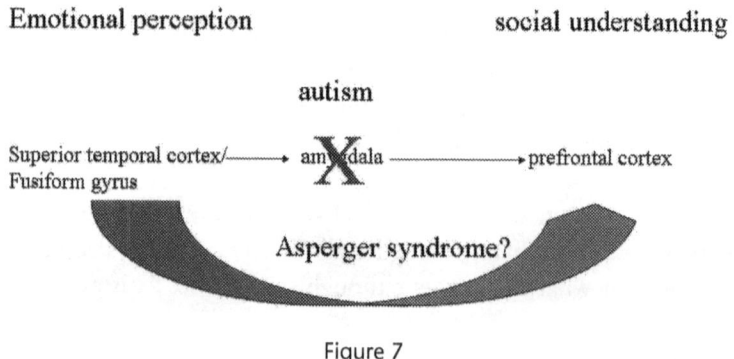

Figure 7

A common misconception is that autistic individuals lack emotions. Many people picture a mute, asocial individual sitting in the corner, rocking back and forth, without any discernable affect. This is simply not true. On the contrary, autistic individuals are hyper-emotional. They certainly laugh and cry, scream with temper tantrums, and often do enjoy the company of others. It is not so much the quantity of emotional expression that is pathological, but rather its awkward quality that characterizes autism. Autistic people will abruptly shift from gushing delight to raging anger to detached calm within minutes, without any obvious trigger. This may be due to a lack of cortical modulation of the limbic system. The higher feeling and cognitive centers of the PFC are not active enough or perhaps too disconnected to inhibit an overactive amygdala. The resulting frontal disinhibition leads to labile mood swings and anxiety.

Anxiety may be subconscious; simply the result of a tonically active amgydala. At another level, it may be a conscious reaction to, and an attempt to cope with, feelings of confusion or uncertainty. Conscious awareness of an emotionally aroused body can itself activate further anxiety. Paradoxically, anxiety may be a greater problem for those on the higher functioning side of autism. They are more aware of what they're going through, but they also have more knowledge about what is socially expected. The pressure to 'act normal' generates anxiety, which feeds on itself. However, this anxiety in the right doses can simultaneously motivate 'proper behavior'. AS people are anxious to act normal, yet by pretending to be normal they become more anxious.

◆   ◆   ◆

My heart is pounding less rapidly as I walk into the subway tunnel. I can think more clearly now. I knew this would happen. It always does. Why, dammit, why? What's wrong with me? I could kick myself. I try to retrace the events of the night in my mind on the train ride back to Kendall Square. I should have stayed at home or perhaps gone to a frat party with some people I knew. I should have tried to make more normal conversation with those Harvard girls. I should have approached those guys from my dorm instead of running away from them. A million maybes. Everything makes so much more sense on reflection. Social faux pas are much less damaging when seen in perspective, and could be avoided if I acted more calmly. But I never do in the heat of the moment.

5

Contingency

Rock and roll will save my soul!
Save my soul with rock and roll!
(WHOOHOO!!)
Rock and roll will save my soul!
Save my soul with rock and roll!
(WHOOHOO!!)...

In the fifth grade, I had a big collection of science fiction and space travel memorabilia: hundreds of newspaper clippings of manned and unmanned NASA missions, science fiction novels, Star Wars bubble gum cards, and so on. I hadn't actually read most of the science fiction, but I really liked the cover art depicting intergalactic spacecraft and alien civilizations. I spent hours each day rummaging through and categorizing this messy stash in my big desk drawer, lost in pleasant concentration. I would routinely ignore calls for dinner and neglect my homework. One day, when I came home from school, I discovered that my father had thrown the entire collection into the garbage. He yelled at me to stop wasting my time and concentrate on my schoolwork. Blinded by tears of rage and loss, I screamed and screamed at my father, banged my head into the wall, and pushed away my anguished mother when she came to console me. For days, I engaged in strange behavior: repeatedly clearing my throat seven times (because seven is a 'complete' number), telling my mother in Korean, "the circle goes around and around...the circle goes around and around..." in response to any statement or question I found annoying. I did eventually get over that particular episode, but there were many more.

My grandfather used to say that I was a difficult child on account of my very strong 'sense of self.' But he added that although I was self centered and stubborn, I was usually not malicious, devious, or vindictive. Indeed, I was often the victim of malice from schoolmates. After a day of being bullied and teased, I would passively build up a mental record of 'injustice points.' Mean comments like 'ching chong', 'Hong Kong phooey', or 'go back to where you came from' usually got one or two points, while physical abuse like pushing or messing up my hair warranted three or more points. After coming home, I would have to redress the point imbalance by taking it out on my innocent mother or on the furniture. I would take a knife and poke seven holes in my bed, and then yell "Fuck You!" seven times until I felt calm and 'balanced'. Eventually my parents had to discard that 'holey mattress.'

Like my father, I have always had great difficulty controlling my temper. It can be provoked by what to others may seem insignificant. Sometimes I'm passive even in the face of great provocation. Other times I get ticked off by the stupidest little thing. I'm not sure why. But once aroused, I find it difficult to back down from conflict, especially when I feel that some 'principle' has been violated. Never mind that this principle may be totally obscure or bizarre to anyone other than myself. This self-righteousness has, unfortunately, led me into some ugly and unnecessary confrontations.

Several years ago, when I was living on the Upper West Side of Manhattan, I used to take the cross-town bus through Central Park to get to my job at Lenox Hill Hospital on the Upper East Side. One day, I was sitting with some heavy bags in an aisle seat next to an elderly woman on a crowded bus. A block before my stop, the woman said, 'excuse me, I'm getting off.' Not wanting to get up quite yet with my heavy load, I told her that I too was getting off at the next stop and continued to sit while the bus crossed the intersection. She then turned to me and said, 'but I'm getting off now! That's the difference!' I could see several other riders turn in our direction, perhaps sensing the rising tension. I felt a surge of mixed emotions: embarrassment, anger, and excitement, well up uncontrollably into my throat. Suddenly, at the top of my voice, I yelled, 'OK, LADY, I'LL STAND UP RIGHT NOW!!' I stunned myself. Immediately, I felt a powerful wave of self-righteous relief, followed a moment later by self-conscious shame. I should have known better.

In addition to my short temper, I have difficulty with motivation. Left on my own, I could spend hours wasting my time with thoughts about military history or the meaning (or meaninglessness) of life. I perform much better in more organized, coercive environments. For example, I did well in high school, where the day was broken down into 55-minute periods and each subject was taught by a familiar teacher in a familiar classroom day after day, week after week. The home-work was routine and predictable; all I had to do was stay a step ahead of the other kids and a chapter ahead of the teacher. My father's beatings provided extra incentive.

I fared badly in unstructured environments such as recess, lunchtime, gym, and school trips. After I came home from school each day, I would think about how things went wrong, and all the bullying I suffered, and how I could make the next day better. But the next day would always be different. One day, it was the big, menacing Bobby Mowen and his friends locking me up in the bathroom and repeatedly asking me if I was 'a chink'. The next day, it was the pretty blonde girl sweetly asking if she could copy my homework. I mishandled both situations. In a predictable environment I would shine, but the real world is never that predict-able.

Meeting new people is stressful. I deal with social demands by organizing and rehearsing mental (and sometimes written) checklists of dos and don'ts. 'Shake their hand and make eye contact; don't talk too much about yourself; repeat their names in conversation, so you won't forget them', and so on. By conscious effort, I can make my behavior seem spontaneous. But my default mental state is one of inertia. I don't usually think of new things to say or new ideas to try without some compelling impetus. In addition, my behavior is captured and guided by the 'here and now.'

Let's say I'm surrounded by people at a cocktail party. I smile while holding a glass of wine. I try to think of something to say, something witty and inoffensive that won't make me look foolish. Then suddenly, I become self-conscious about standing there saying things that I don't really mean to someone that I don't really care much about. As often as this feeling occurs, I'm always stunned by it, and find it difficult to continue the initiative in conversation from that point on. I finish the sentence, smile politely, and let the other party continue, while I anx-iously await my cue to go home. This is a feeling I will call 'intentional disso-nance'.

Strangely enough, I sometimes experience intentional dissonance in solitude as well. I may be cleaning the apartment or getting ready for work when I am suddenly overcome by an overpowering feeling of lassitude. I usually give in to the inertia and lose my will to do anything productive. These quasi-catatonic states can last for hours, especially when I'm tired. But they are not unpleasant. Indeed, some of my happiest moments occur during these spells. Ultimately, however, I must return to reality. I curse myself for having wasted hours doing nothing or flipping through the television remote when I could have been making new friends.

How much are the flaws of impulsivity, inflexibility, and indecisiveness parts of my true self? On one hand, I am very well aware that they are flaws. If I were to come upon a magic vase from which a genie arose to grant me three wishes, I might be tempted to say, 'first, grant me the strength to change the things I can, second, the courage to accept the things I can't, and third, the wisdom to know the difference.' I do want to be more patient, more even tempered, better able to fix things with my hands and figure out new things with my mind, to be able to cook a four course meal while taking care of children and getting to work on time the next day. But my 'true' desires often seem to be derailed by unpredictable and irrational forces. Only by a deliberate exercise of will can I consistently take control. On the other hand, when I do force myself to do and say the right things, it doesn't feel natural. It would certainly be nice to be able to multitask, come up with creative decisions at the annual board meeting, or make friends at every cocktail party, but would it still be me?

Executive Functions

Flexibility, restraint, and decisiveness are universally regarded virtues. The individual who possesses them to a high degree is described as mature and refined. Why are these characteristics considered to be desirable, and what do they have in common? Loosely defined, they are what neuropsychologists call the **'executive functions'** (**EFs**). Along with language, the EFs are the most highly evolved mental attributes in the animal kingdom. They are what separates and elevates human intelligence from all others. When we walk down an aisle in the grocery store and resist the urge to buy ten boxes of ice cream, or decide to take a new shortcut to work, we are exercising 'executive control'.

EFs are mental processes that enable an individual to control her or his attention, shifting it to some aspect of the outside world (a moving target, a familiar face, the smell of roses) or internal representation (next year's projected tax rebate, the memory of a tequila hangover) in order to make judgments and solve problems. EFs integrate the cognitive machinery of sensory perception, emotional states, creative imagery and memories to form a coherent stream of thought which can then be used to guide action.

Without proper executive control, complex behavior becomes impossible to coordinate, and actions and thoughts become disjointed. What would it be like to have no executive control? Perhaps we would be like a scientist who could never finish an experiment because she would have to start from scratch after every interruption, or a child who stops his mother in front of every store window on the way to school. Autistic people have these characteristics. In this chapter, I will argue that deficits in EF and the circuits underlying them in the brain may explain much of what is central in the autistic disorders.

◆   ◆   ◆

Humans and other animals have evolved wonderfully efficient bodies and brains designed to work within their natural habitats. Just as our four-chambered heart beautifully regulates the volume of blood coursing through our lungs and the endocrine system fine-tunes the release of hormones depending on the concentration of sugar and salt in our blood, the brain is modeled to sense and respond to its external environment. But the environment that the brain senses is much more complex than blood pressure or blood sugar levels. Rather, it is an environment populated by complex objects with unpredictable trajectories, strange sounds and smells emanating from unexpected sources, fellow creatures motivated by their own self-interest, and even figments of the imaginative mind.

Much of the environment is regular and predictable: the sun generally does rise in the east, the days do get shorter approaching the winter solstice, mothers usually are affectionate, jumping into a swift river is generally dangerous. The brain has evolved marvelous modules for representing such generalities and handling their contingencies through the deployment of stereotypical behaviors. For example, infants show an innate fear of heights and snakes, and will avoid them. Infants have an innate attachment for their mothers (and mothers for infants), and will cry for them. But the social world that we inhibit after childhood is far more complex. It demands creative strategies for dealing with vague or conflicting goals.

Discovering what one's goals are at a particular time, formulating a workable plan for attaining them, coming up with alternative strategies, and moving on to other goals afterwards are tasks that require the executive functions: mental flexibility, concentration, and creativity. EFs are easy to grasp in principle, but trickier to define exactly. One way of looking at them is to divide them into three types: inhibition, flexibility, and planning.

Inhibition refers to the ability to maintain attention without getting distracted by other stimuli. A failure to inhibit leads to impulsiveness, allowing one's behavior and thoughts to be 'captured' by the immediate surroundings rather than by the desired goal. This results in 'prepotent' rather than appropriate responses. The ability to inhibit prepotent responses can be gauged experimentally by the '**Stroop test**'. This clever psychological device, invented by a man named Stroop in the 1930s, involves viewing a list of color words printed in contrasting ink, for example GREEN written in red ink. The subject must identify the colors of the words as quickly as possible. (**figure 1**) Getting a high score on the Stroop test requires the ability to inhibit one's natural tendency to read the word (the more salient stimulus for those who can read) rather than say the color.

control	**test**
RED	GREEN
GREEN	BLUE
RED	RED
BLUE	GREEN
GREEN	RED
RED	BLUE

Figure 1: Stroop test

Flexibility is another EF. In contrast to inhibition, it involves the shifting of attention from one thought or action to another. This is called 'set-shifting'. Failure to shift attention leads to perseveration, the repetition of a prior thought, action, or action component in a nonproductive manner. Repetition of sounds, words, or phrases, narrow circumscribed topics of interest and conversation, and resistance to change are all examples of inflexibility at different levels. A widely used neuropsychological test of flexibility is the 'Wisconsin Card Sort Test' or WCST. This exercise requires the subject to sort through a special deck of cards by shape (circle, square, triangle, cross), color (red, blue, green, or yellow), or number (one, two, three, or four). The subject is not told which is the proper sorting protocol, but only whether a choice is correct or incorrect. After ten consecutive correct choices (implying that the subject has now 'gotten' the rule), the rule is abruptly changed (without telling the subject), and the test proceeds. The subject is scored on the number of cards correctly sorted.

The third major executive function involves the planning and monitoring of the decision making process. This requires the ability to hold goal-directed information 'on-line' while other components or sub-components of the unfolding action are attended to. Take the relatively simple act of getting up in the middle of the night for a glass of water. The task involves a complex sequence of discrete action components that must be followed: getting out of bed, walking to the kitchen, turning on the light, getting the glass from the cabinet, turning on the cold water tap, and so on. Each component has sub components as well: for example 'getting out of bed' involves opening the eyes, pulling the sheet off, lifting the neck, pushing off with the arm, extending the leg off the bed, and so on. Each sub component has sub sub components that are ultimately reducible to individual contractions of muscle fibers triggered by motor nerve impulses.

Keeping the main goal in mind while the myriad sub goals are attended to in proper order is accomplished through the 'working memory' system. Working memory at its most basic level simply refers to the maintenance of a goal-directed mental representation of an object, location, or relationship through a variable (several seconds) delay period. Failure of working memory results in a poverty of action. Goals and/or the routines through which they are realized are poorly represented. As a result, the individual is constantly mired in hesitation and indecision. She may not know what she wants, or know but be unable to formulate a plan to attain it. Planning ability can be assessed by tests of working memory such as the 'Tower of London' and 'Tower of Hanoi'. In these tests, subjects are required to move a set of colored rings one at a time from one peg to another in the fewest number of steps.

Executive Hierarchy

Executive functions are not all created equal. Simple well-learned or 'over learned' actions such as getting a glass of water or driving to work are governed by a lower level of EFs than something more difficult, novel, or unnatural such as playing chess, initially learning to drive, or rushing into a burning building to save your child. Set shifting, inhibition, and working memory are all as important as before, but the desired goal cannot be attained by a series of preset automatic steps. Rather, there are multiple possible sequences of steps competing for attention: 1 d4 d5 2 c4 c6 3 Nc3 Nf6 4 e3 e6 5 Nf3 Nbd7 6 Qc2 Bd6 7 g4 dc 8 Bc4 b6 9 e4 e5 10 g5 Nh5 11 Be3 0-0 12 0-0-0 Qc7 13 d5 b5 14 dc bc 15 Nb5 Qc6 16 Nd6 Bb7 17 Qc3 Rae8 18 Ne8 Re8 19 Rhe1 Qb5 20 Nd2 Rc8 21 Kb1 Nf8 22 Ka1 Ng6 23 Rc1 Ba6 24 b3 cb 25 Qb3 Ra8 26 Qb5 Bb5 27 Rc7 resigns (former world chess champion Gary Kasparov defeats the Deep Junior supercomputer, 1999); flicking on the left blinker, checking the rearview, turning the steering wheel to the left, and stepping on the accelerator to pass vs stepping on the accelerator, flicking the highbeams several times, and lightly braking three feet behind the car ahead; running back into the fire vs dialing 911 again. It is not at all clear which combination would best achieve the goal, or even what the goal may be at any given time. The psychologists Tim Shallice and Donald Norman have proposed an influential theory of executive control called the 'supervisory attentional system' model (SAS), which goes some way towards explaining how automatic fragments or 'quanta' of behavior may be organized into coherent, conscious goal-directed action. (**figure 2**)

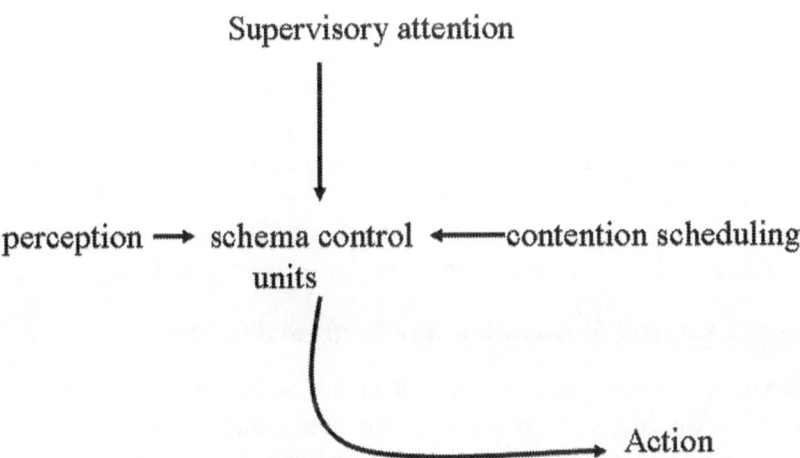

Figure 2: the supervisory attention system

According to Norman and Shallice, all action, voluntary or involuntary, conscious or subconscious, can be subdivided into small components which they call 'action schema'. A series of action schemas combine to produce an act. The choice of schema actually selected and activated depends both on perceptual input and on competition among the various possible schema units for activation. Schema units, like units of perception, are thought to reside in discrete nodes within complex neural networks. These nodes have a variable level of activation that depends on how well each is connected to its neighbors. The connection 'strength' between the nodes may be preset by genes, affected by learning, and further modified by higher supervisory influences. For instance, infants are pre-wired to suck when placed in proximity to their mother's nipple (or a pacifier or finger). We learn to cover our mouths when we yawn. If two schemas are habitually associated with each other, the connection between their nodes is strengthened; if not, the connection is weakened. Additionally, the activation of one node will tend to inhibit all surrounding non-linked nodes. Sensory/perceptual input activates the action schema most strongly wired to it, which in turn stimulates those schemas most strongly associated further down the line, eventually producing the final action.

If this were all there was to human behavior, we would be little different from cats and dogs, acting largely on impulsive instincts and maybe some simple learned associations between various stimuli. There would be little in the way of planning, self-control, flexibility, creativity, and, indeed, consciousness as we know it. No complex action could be carried out; no complex thought could be entertained.

But there are many occasions that warrant more flexible behavior. This requires a **supervisory attentional system** (SAS). The SAS modulates the activation values of the various action schemas to facilitate a more appropriate response. It is generated by a part of the brain that is well connected to all the sensory, emotional, and memory circuits as well as to the various lower EF circuits that mediate a motor response. In addition, it performs a continuous monitoring function that modifies the representation of goals in real time.

Executive Control & Attention: the Prefrontal Cortex

Executive functions require attention. To inhibit prepotent responses, such as those that occur during the Stroop test, one must maintain attention on the requisite task while inhibiting attention to the distractor. Flexibility and set-shifting, as in the WCST, involve an efficient disengagement mechanism that shifts attention from one location to another. The SAS requires attention to be directed first to the main goal, and then to the sub-goals necessary for its successful completion.

In order for the SAS to work, all potential objects of attention must be held on-line for a period of time while it is recruited. There is ample evidence from both brain injury patients and monkey studies that this function occurs in the prefrontal cortex. Damage to the PFC causes impulsivity, perseveration of thoughts and behavior, and poor planning of routine daily activities. Tim Shallice has found that a group of frontally injured patients with otherwise good memory and intelligence failed miserably when told to carry out routine errands such as going to the store to purchase a newspaper. These patients also scored poorly in neuropsychological tests of EF, such as the Stroop, WCST, and Tower of Hanoi.

Neuroscientist Patricia Goldman-Rakic has done a series of experiments on monkeys trained to find targets after delays or distracters. These experimental monkeys were briefly shown a visual cue, such as a red dot or blue cross, which they would learn to associate with a food reward at one of several covered wells. They would then be distracted for a time before being allowed to select the proper well (containing the food). Goldman-Rakic found neurons in the monkey prefrontal cortex which are active during the delay phase of these so-called 'working memory matching tasks'. These and other animal experiments have revealed that the PFC functions to form dynamic neural ensembles that link the sensory association areas (the sites of higher sensory perception, memories, and imagery) and the limbic structures (mediating emotions) with the sites of motor output. In effect, the ensembles **represent** the associations between objects, feelings, and locations relevant to the goal, and modulate motor areas to influence behavior until the goal is met or another 'more urgent' goal is selected. The neurons in the PFC are unlike those in the sensory cortex in that they code not a particular stimulus, such as a line segment in space or the pitch of a musical note, but rather **the specific contingencies of a task to be accomplished**.

A final note should be made concerning the concept of 'motivation'. Simply put, motivation is what makes us want to do something. It directs behavior towards bridging the distance between the present state of body and mind, and an imaginary state in which current goals and desires are satisfied. Motivation is the driving force that directs executive function to do what we want to do.

But what does this really mean? Motivation is based on a neurochemical reward system that suffuses the prefrontal executive circuits whenever goals are satisfied. The 'rewarding' neurochemical in question is dopamine. Dopamine is released from the midbrain after winning a race, getting promoted, or having a drink with friends. In fact, virtually all drugs of abuse from nicotine to heroin to crack cocaine work by enhancing the dopamine effect in the central nervous system. Dopamine enhances the activity and connectivity of the prefrontal neurons to the other brain regions mediating attention to strengthen the network that led to its release. The next time you find yourself in the same situation, the acclimatized PFC will know exactly what to do. It will select the proper action schemas to enable prompt, efficient, and definitive decision-making. (**figure 3**)

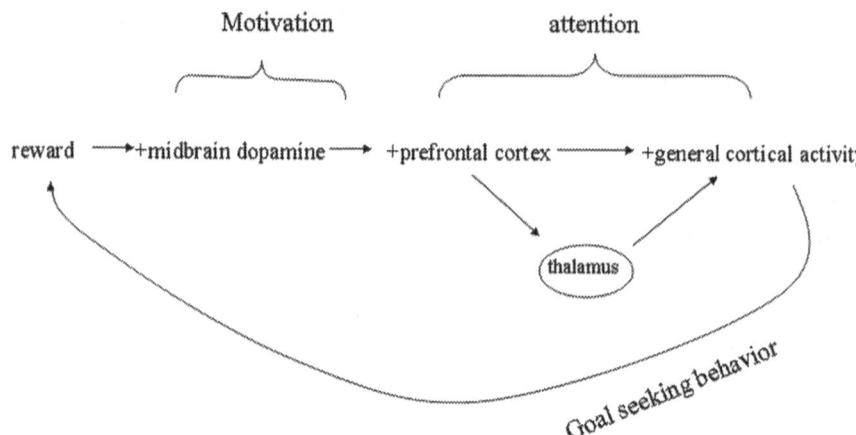

Figure 3: motivation and attention

Free Will: the Anterior Cingulate

We have seen that attention is mediated by highly selective and dynamic neural ensembles in the PFC. The motivating force behind attentional control is the 'reward signal' of dopamine released into the prefrontal networks, strengthening or weakening particular ensembles. At the apex of executive control is the SAS, the mechanism for error detection and conflict resolution. One part of the PFC appears to be the physical locus for the SAS: the **anterior cingulate cortex**.

Several neuroimaging studies have found that the anterior cingulate (AC) area is activated by novel situations, inhibition of prepotent responses, error detection, and conflict monitoring. In one intriguing study, subjects were stimulated in one finger of the right hand and told to move that finger while having their brain activity analyzed by a PET scanner (the control condition). In the test or 'free will' condition, they were stimulated as before, but told to move a random finger. When the PET activities of the two conditions were compared, it was found that the free-will condition was associated with a significant increase in AC brain activity.

Bilateral damage to the AC sometimes causes a curious disorder called 'akinetic mutism'. This is how Antonio Damasio describes a woman so afflicted:

'She suddenly became motionless and speechless, and she would lie in bed with her eyes open but with a blank facial expression…Her body was no more animated than her face. She might make a normal movement with arm and hand, to pull her bed covers for instance, but in general, her limbs were in repose. When asked about her situation, she usually would remain silent, although after much coaxing she might say her name, or the names of her husband and children, or the name of the town where she lived. But she would not tell you about her medical history, past or present, and she could not describe the events leading to her admission to the hospital. There was no way of knowing, then, whether she had no recollection of those events or whether she had a recollection but was unwilling or unable to talk about it. She never became upset with my insistent questioning, never showed a flicker of worry about herself or anything else. Months later, as she gradually emerged from this state of mutism and akinesia (lack of movement), and began to answer questions, she would clarify the mystery of the state of mind. Contrary to what one might have thought, her mind had not been imprisoned in the jail of her immobility. Instead it appeared that there had not been much mind at all, no real thinking or reasoning. The passivity in her face and body was the appropriate reflection of her lack of mental animation. At this later date she was certain about not having felt anguished by the absence of communication. Nothing had forced her not to speak her mind. Rather, as she recalled, 'I really had nothing to say.' [**Descartes' Error** pg 73]

Akinetic mutism is not to be confused with the so-called 'Locked-in syndrome' caused by a stroke in a part of the brainstem region. This syndrome causes paralysis of all facial and bodily movement except for the eyes, but appears to leave awareness, cognition and 'intention' intact. Patients in this horrifying state are bed bound, on respirators, and are unable to communicate their claustrophobic anguish other than through feeble eye movements.

The results of neuroimaging studies in 'free will' conditions and in patients with akinetic mutism have prompted Sir Francis Crick to propose, half seriously, that the AC may be the 'seat of the human soul'. Indeed, what we commonly refer to as 'free will' may be nothing more than attentive decision-making as mediated by activity in the AC and other prefrontal areas. There is now quite a bit of convincing scientific evidence that activity in the AS actually precedes the conscious intention to act. If so, consciousness itself may be a consequence of attentional brain activity.

Practice makes Perfect

Much of what we do each day is routine. We wake up, get out of bed, drag a comb across our head, and so on without much thought. These actions don't require much in the way of executive control; indeed, it would be a waste of mental resources if we had to exercise attention and willpower in the form of SAS circuits and anterior cingulate activity each time we ordered a cup of coffee. There is also much experimental, as well as anecdotal, evidence that paying attention to actions that are normally performed without conscious thought causes a deterioration in performance. Professional athletes and actors are familiar with this phenomenon.

Just as memories are slowly transferred from the hippocampus to the cortex (more on this in the next chapter), learned actions are gradually transferred from the cortex to the basal ganglia (BG) as they become habitual. Recall that the BG form parallel circuit loops with the cortex. Specific motor or cognitive functions in the cortex, such as the ability to whistle 'Dixie', integrate 'sine x', or recall the first thirty lines of The Rime of the Ancient Mariner, are 'mapped' onto a corresponding area of the BG. **Motor routines** appear to be preferentially mapped to a part of the BG called the **putamen**, while **cognitive routines** are mapped to the **caudate nucleus**. (**figure 4**)

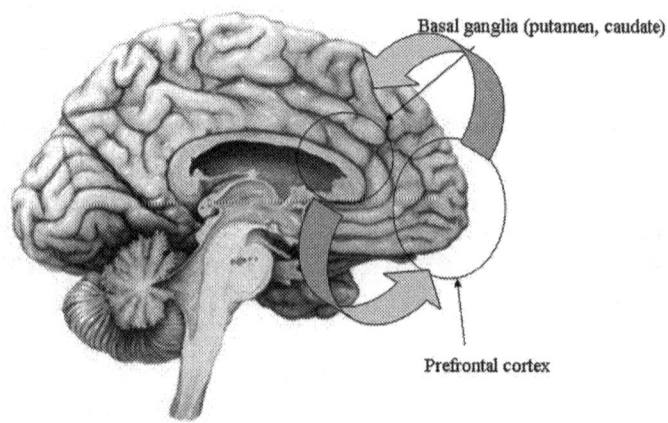

Figure 4: cortico-basal ganglia loops

The BG are important in the establishment of learned routines and habits. They relieve the cortex from having to attend to mundane activity. Additionally, they integrate the components of habits into a smooth operation. The neuroscientist Ann Graybiel has proposed that the BG recodes cortical inputs into 'chunks' of sequential behavior. Thus, once we have learned to swim, we don't have to think, 'left arm stroke, right arm stroke, head turn, BREATHE!...' We just 'swim' or go with the flow without having to think about it at all. In this sense, the output of cortical-BG loops is analogous to the action schemas in the

SAS model: once a loop is selected, the entire ensemble is expressed as a chunk. Normally, we have control over the deployment of these automatic loops. We can decide when to swim or stop swimming, how fast, in which direction, and so on. These are the top-down attentional signals from the PFC. But there are two well known disorders in which conscious control of habitual behavior is derailed: **Tourette's syndrome** (TS) and **Obsessive–compulsive disorder** (OCD).

TS is characterized by repeated 'tics': sudden movements or vocalizations that resemble fragments of purposeful behavior. People with TS may suddenly jerk their arms, clear their throats, or even utter obscenities. These are not willed actions; the individual is helpless to suppress the urge to tic, which is exacerbated by stress and fatigue. Although this urge can be voluntarily suppressed for periods of time, they must ultimately be released as 'a swimmer needs to come up for air', as one TS sufferer put it.

◆   ◆   ◆

OCD and its cousins, obsessive-compulsive personality trait, hypochondrias, and the various body dysmorphic disorders such as anorexia nervosa and bulimia, are characterized by unwanted thoughts (obsessions) such as the nagging feeling that one's body or environment is unclean, unhealthy, or somehow 'not right'. To ameliorate these obsessions, the sufferer compulsively repeats ritualistic behaviors to reduce the anxiety (washing one's hands, checking the stove, getting a blood test, bingeing and purging) dozens or hundreds of times.

TS and OCD share some similarities. As many as 90% of TS individuals also have some characteristics of OCD. Both disorders involve urges to engage in repetitive, ritualistic and otherwise unwanted behavior. In both cases, the behavior can be consciously suppressed for some time. The neural link between the disorders is the cortical-basal ganglia loops mentioned earlier. TS is thought to involve dysfunction in the 'motor loop' through the putamen, while OCD involves 'cognitive loops' through the caudate. The normal facilitation of these loops, once again mediated by dopamine transmission, is abnormally up regulated. Chunks of habits and thoughts come rushing to the surface, overwhelming the normal censorship control by the PFC. Effectively, individuals with TS and OCD become stuck doing things and thinking thoughts that are no longer appropriate or adaptive to the contingencies at hand. In these disorders much executive control is lost, leading to perseveration.

Autistic Repetition: Failure to Escape

Repetitive and perseverative thoughts and actions are extremely common in the autistic spectrum disorders. They span a wide range of behaviors from simple motor actions to elaborate routines and all-encompassing interests. We can think of repetitive behavior in a hierarchical manner. (**figure 5**) At the bottom are motor behaviors akin to the tics of TS: grimaces, grunts, spasms of eye blinking and arm jerking. In addition, some autistics, commonly those with low IQs, display stereotypical body rocking and head banging. Repeated behaviors can range from simple action components consisting of isolated muscle contractions to more complex sequences of movement that begin to approximate purposeful activity.

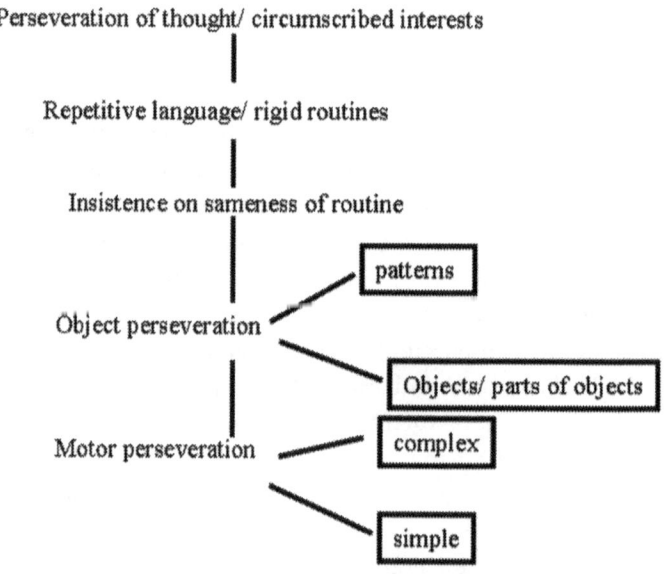

Figure 5: repetitive hierarchy

In 'self-perseveration', the object of repetitive action is a part of one's body. Above this is 'object perseveration', repetition upon an 'other' object. Here, too, there are multiple 'levels'. Some children are consistently preoccupied with certain parts of things such as spinning rotors or shiny wires. They may collect and tinker with electric motors and circuit boards to the exclusion of other toys and activities. Other autistic individuals may be fixated on patterns of objects. They spend hours organizing shoes, books, toys, or Chinese take-out menus into neat rows and stacks. The level of organization may be different, but they share an insistence on repetition. Insistence on sameness of the environment is not a repetitive behavior, but it does involve inflexibility in thought. Autistic individuals may become very upset if a well-meaning person reorganizes the 'clutter' on their desk or cleans out their bedroom closet. The insistence on sameness of things (in space) can also apply to actions (in time), resulting in rigid adherence to routines and rituals. For example, one may 'have to' see a certain television program, no matter what, or read the morning paper in a certain order, or go shopping before working out in the gym. If the program is cancelled, or the papers are disordered, or the store is closed, the result is inconsolable rage and inertia; the rest of the day may be ruined. Similar to rigidity of routine is inflexibility of language. Autistics often demonstrate perseverative speech, asking the same question repeatedly or echoing what they've just heard. We will return to the peculiarities of autistic language in chapter 8.

Finally, at the top of the repetitive hierarchy are circumscribed interests. This characteristic represents a perseveration of conscious thought, rather than of subconscious action. Nonetheless, the repetitive motif is preserved. Autistics can go on and on about the colonies of the British Empire, major league batting averages, or makes and models of toaster ovens. The particular interest can vary widely depending on the intelligence, education, and personality of the autistic individual. In addition, a single individual may pass through many different interests. For example, I was obsessed with space travel at age 10, British History at age 18, and neuroscience at age 30. I currently have no more than a passing interest in rocket ships and science fiction. Perhaps it is this inflexibility of autistic thought, especially among the more intelligent and well-read individuals with Asperger syndrome, that forms the basis of their often dogmatic beliefs and stubborn character.

What all these repetitive behaviors have in common is their detachment from the contingencies of goal-directed action. Autistic perseveration, whether involving finger flicking or an obsessive interest in New Yorker cartoons, is maladapted to dealing with the here and now. Autistics indulge in tics and thoughts not as a means to some end, but as ends in and of themselves.

Earlier theorists of autistic psychology speculated that repetitive behaviors may be subconscious 'coping mechanisms' to deal with the confusion of sensory overload, especially in socially demanding situations, much as 'neurotypicals' might engage in a night of repetitive drinking and dancing to blow off a stressful work week. But this explanation can't be right because most autistic people tend to be even more perseverative and preoccupied with their routines when in less socially stimulating environments. In fact, AS individuals, like TS sufferers with their tics, can suppress their perseverative urges quite well in public.

I believe that autistic repetition results from defects in executive control of behavior and thought, especially attention shifting. Flexibility in both motor and cognitive control involves the ability to shift attention from one neural representation (of a muscle group, body image, action schema, visual target, verbal construct, moral standard) to another. Without proper and timely shifting based on sensory and motivational input, behavior cannot be guided to deal with external contingencies. Rather, it is trapped in a perseverative loop; there is a failure of behavioral 'escape'. There is strong evidence supporting failure of attention shift in autism. Not only are autistics impaired in real world tasks of flexibility, but they consistently fail set-shifting tests such as the WCST relative to IQ matched control subjects.

It is interesting to compare autism with OCD and TS. Repetitive and stereotypical behaviors are even more prominent in these disorders than they are in autism, yet experiments seem to suggest that there is less failure of set shifting (although better controlled head-to-head comparisons with autistic groups have yet to be undertaken). Autism might involve a different neurochemical or anatomical defect. Research in TS and OCD is beginning to focus on hyperactive cortical-BG loops which function independently of PFC modulation. Imaging studies have indeed shown increased activity in various regions of the BG in TS and OCD. In autistic subjects, no such abnormalities have been found, but there is evidence of blood flow abnormalities in the PFC. Could it be that the repetitive behavior of TS and OCD is due to overactivity of the BG, while that of the autistic spectrum disorders is related to reduced activity of the PFC, with both leading to similar outcomes through very different routes?

◆   ◆   ◆

It is instructive to contrast autism with another developmental disorder, **Attention Deficit Hyperactivity Disorder**, or **ADHD** for short. ADHD, usually diagnosed (or overdiagnosed in the United States) in grade school boys with behavioral and learning problems, is characterized by an inability to maintain focused attention, leading to impulsive, hyperactive behavior. ADHD has been found to be associated with increased dopamine activity in the PFC and the anterior cingulate cortex. Perhaps here lies the difference between the failure of inhibition in ADHD and its relative sparing in autism: frontal hyperactivity causes increased 'attentional escape' or shifting, while reduced activity produces less escape and more capture (perseveration). This model is likely to be oversimplified, though. Many autistic individuals, myself included, do show some degree of impulsiveness.

Autistic Inertia: Failure to Generate

Finally, we come to the highest of the EFs. Planning and decision making, along with the ongoing monitoring of activity, allows us to generate or imagine goal-directed behavior that is both focused and flexible. The commonly used lab tests for the ability to plan, such as the Tower of Hanoi, have been conducted on autistic subjects in a number of studies. Most of them have significant deficits compared with controls. In addition, psychologist Michelle Turner has found that they perform poorly on tests of creative or generative ability. Examples of these tests include giving a picture of common objects (garden hose, mop top, glockenspiel) for which the autistic subject must come up with a maximum number of creative uses, or a sheet of blank paper on which the subject is told to draw as many designs of just four lines as possible in a minute. Autistic subjects tended not only to perseverate and repeat responses, but also demonstrated poverty in the total number of responses. It suggests that planning and generation of new responses, as well as attention shifting, are lacking. This may explain the lack of imagination common in autism.

Theory of Mind versus Executive Function

There is currently some lively debate in academic circles concerning whether EF, TOM, or central coherence is 'the primary problem' in autism. Simon Baron-Cohen supports the existence of a defective TOM module separate from, and perhaps responsible for, the EF deficits found in autism. James Russell, Sally Ozonoff, and others believe that many, and perhaps all, of the so-called TOM deficits can be explained by underlying EF problems. Uta Frith has advanced the central coherence theory that attempts to account for many of the EF and TOM deficits in terms of a failure to integrate lower levels of representational information into more 'coherent' wholes.

An interesting paper examined a patient with concomitant schizophrenia and Asperger syndrome, presumably secondary to congenital damage to his left amygdala [**Fine, C., et al, Dissociation between 'theory of mind' and executive functions in a patient with early left amygdala damage', (2001) <u>Brain</u>, 124, 287-298**]. The subject was found to score poorly on all first-order and second-order TOM tests (such as the Henry and Victor test described in chapter 3), but did well on all tests of EF (WCST, Stroop, Tower of London, etc). These results, plus reports of frontal brain injury patients found to have intact TOM with grossly impaired EFs, have led the authors to write, '…the findings clearly suggest that theory of mind is neither mediated by nor necessary for executive functioning. Rather, the present findings suggest that theory of mind is mediated by a…dedicated neural system.' [Fine, pg 295] However, such dissociations of TOM and EFs are unusual; these deficits **usually** occur together in both acquired brain damage and developmental disorders such as autism. TOM and EFs may be different, but interrelated entities.

The central coherence theory takes the middle ground: autistics cannot understand the intentions and beliefs of others because they cannot construct a metarepresentation of others' mental states. Similarly, they have difficulty with planning and shifting focus because they cannot integrate the bits and pieces of the sensory/perceptual stream into a purposeful whole. Here lies the appeal of the central coherence theory. It depicts EFs and TOM as composite mental modules that build different types of coherence into the world. EFs build **general coherence**, based on the laws of probability, similarity, and causality—in short, the world of folk physics. TOM builds **special coherence**, based on understanding others as free agents—the world of folk psychology. What the central coherence theory does not describe are the units that make up the composite modules: the individual quanta of coherence. These, I propose, are attentional circuits.

Alan Leslie, who, along with Uta Frith and Baron-Cohen, is one of the founders of the TOM account of autism, has recently proposed a new twist on the old idea. He suggests that the abstraction mechanism which enables TOM is itself a form of attention shifting. We create beliefs about ourselves and about others by shifting attention from agent to agent. TOM, like EFs, is a way of paying attention to and shifting attention from bits of information that we ascribe to 'others', be they people, minds, things, events, ideas, or contingencies, to serve our biological and ecological needs.

Intentional Dissonance

When the attention circuit is working well, there is a continuous loop that smoothly resonates from sensory systems to motor systems in the brain. Some loops are shuttled through various areas that subserve emotional information, memories, internal images, and habitual responses. The PFC, including the anterior cingulate cortex, monitors this ongoing activity, sometimes sitting back, sometimes actively taking the reins. There is a Zen-like harmony and balance between observation and action, passivity and activity. It is an 'intentional resonance'. But in some cases, due to a genetic mistake, developmental glitch, biochemical imbalance, toxic insult, or some combination of the above, the resonance is disturbed. Executive, attentional monitoring of the loop is lost, resulting in a sort of 'intentional dissonance'.

It is curious that there are times when I become very self-conscious of myself in **the act of acting**. I am attending not to the act, but to myself, almost as if I am looking down at a stranger in my place. Perhaps this is why I prefer to watch others engage in the games of life rather than to throw myself into the fray. It feels better to **be** than to **do**.

6

Geek

It was another humid afternoon in the summer of 1979. There was a hint of a salt breeze coming off the Gulf Coast a mile to our south, but the air was still heavy and sullen. The hum of the cicadas filled our ears. I was looking forward to the next subject for the afternoon, earth science, but no one else seemed very interested. Most of my classmates in the fifth grade at Gorenflo Elementary School in Biloxi, Mississippi were children of shrimp fishermen or military brats like me stationed at nearby Keesler Air Force Base. I was a child of Korean parents; the rest of my class were black, white, and Vietnamese refugees. This was one of the reasons I was different. But I also stood out in another way. I was 'smart'.

Mrs. Lenham asked the class to name the layers inside the earth. I raised my hand immediately. Everyone looked at me. 'Does anyone besides Henry know the answer?' she asked. No one else seemed to know or care. After a moment she asked me to share my knowledge.

'The crust is the outside layer; it's about three to fifteen miles thick depending on whether it's under the ocean or under mountains. Under that is the mantle. It's made of dense rock and is about 1500 miles thick. The crust forms continental plates that float on top of the mantle. Under that is the outer core, which is molten metal and about 1000 miles thick. And in the center of the earth is the inner core, which is made of solid iron and is about 1600 miles in diameter.'

My classmates must have eyed me with a combination of envy and exasperation. This was pretty much the pattern throughout my grade school years. It seemed like I knew just about everything. My nicknames included 'Mr. Professor', 'brainiac', 'geek', and 'nerd' (in addition to the usual assortment of 'ching-chong', 'gook', and 'chink').

By the time I was 11, I knew the capitals and populations of almost all the world's nations, the dates and casualty figures of most of the major battles of the

Second World War, the names and dates of all the American presidents and British monarchs. I was the little Asian boy with innocent manners, oversize glasses, and a 'rice bowl haircut' who loved to babble on and on about the density of neutron stars (a teaspoon full of their matter would weigh thousands of tons on earth) and life in Cambrian seas (full of trilobites and ammonites) rather than play with other kids my age. Most likely the adults and teachers found me a little strange and awkward. Perhaps they wouldn't have wanted such a child themselves, but enjoyed my company in small doses. I sensed their approval and gravitated towards them rather than to my peers, who generally bullied or ignored me.

Lists have always had a special salience for me: largest cities, longest rivers, bloodiest battles, highest grossing movies, fastest planes, the moons of Jupiter, geologic periods, Wimbledon tennis championships. I felt almost compelled to memorize these things. People would ask me how long I would have to study to learn all this information, but I never really had to work at it. I spent a lot of time after school browsing through the <u>Guinness Book of World Records</u>, the world atlas, and various encyclopedias and almanacs, but all of this was fun, not homework.

Lists of things in time and space were always fascinating to me. To this day, one of my favorite books is a (somewhat subjective) list of the 100 most influential people in history by Michael Hart. I have endlessly pondered the counterfactual historical arguments behind the author's listing of Mohammed (first) and Isaac Newton (second) ahead of Jesus Christ (third). I am also pretty good at learning 'serial processes' in physics, engineering, biology, and history. I can recall, for example, the excitement of learning the trajectory of a Saturn V rocket as it jettisons its stages en route to the moon, the process of transcription and translation of DNA into protein, the route of Alexander the Great's conquests through Central Asia. More abstract non-serial concepts such as grammar and calculus were harder to grasp, but I did well enough in those subjects through lots of hard work. Music and foreign languages, on the other hand, were always difficult for me, and I never did very well in those subjects.

It is interesting that in times of stress or boredom, I find myself silently rehearsing lists like mental tics, sometimes linking them like symbols to actual events and concerns in my life, giving them some bizarre, obscure significance. For example, if I'm reading a rather dull medical textbook of, say, 500 pages, and I'm on page 315 and have another 185 pages to go, I pretend that the pages symbolize the years from 1500 to 2000. Therefore, I'm in the year 1815 (page 315), which, of course, was the year of Waterloo and the start of the second British imperial hegemony, the first 'Empire' having been lost with the American Revolution in 1776 (page 276). If the next three chapters I have to read go from pages 335 to 398, I pretend that those chapters roughly correspond to the Victorian era (1837-1900), and start daydreaming about the glory days of the British Empire. Sometimes I actually forget what I'm reading about.

Thankfully, by the time I was in college I was no longer seen as a child prodigy. First of all, I was no longer a child. Second, as an MIT undergraduate, I was now surrounded by lots of smart people. Third, and perhaps most importantly, my way of accumulating knowledge, as impressive as it may have been in the fifth grade, was no longer adaptive or useful outside of the occasional game of Trivial Pursuit. What is important for success in higher education (and life in general) is the ability to grasp concepts and apply them to real life situations in a timely and flexible manner. I am seriously challenged in this department.

Recently my English girlfriend and I went to see Robert Altman's movie 'Gosford Park'. We both enjoyed it thoroughly, but I think for different reasons. Jane found the plot (an Agatha Christiesque who-done-it set in a 1930s English country estate) and subplots funny and captivating. In contrast, I was quite confused by all the different characters and the dizzy pace of dialogue. But as an incurable anglophile, I anchored the setting of the story to a certain place and time in my historical knowledge base. This allowed me to interpret the movie in a more global socio-political context (class conflict, the decline of Imperialism, American popular culture encroaching on 'high-brow' British society) rather than in more personal terms. I enjoyed the film for what I believed it to mean, and also for its superficially elegant visual detail and English accents. To borrow a phrase from Oliver Sacks, I approach fiction as an 'anthropologist from Mars', being simultaneously fascinated and baffled by the sheer communicative vitality of these intentional creatures we call human beings.

Savants

Most of us are familiar with the autistic savant from the movie, 'Rainman'. The title character Raymond Babbitt (played by Dustin Hoffman as a composite of several real savants) was able to do lightning fast arithmetic in his head, memorize hundreds of license plate numbers, and count cards at the casino. In real life, Stephen Wiltshire, the autistic artist from Oliver Sacks' **Anthropologist from Mars**, drew unbelievably accomplished architectural drawings without any formal training. Tom Bethune, a blind and mentally retarded former slave from

Georgia, had perfect pitch and made a living touring Europe playing Beethoven, Bach, and Chopin on the piano. A pair of identical twins, George and Charles, could tell on what day of the week a given date would fall thousands of years in the past or future. Clara Park, as described in a wonderful book written by her mother, spent months calculating all the factors of 26082, 13041, and 19380 [**Exiting Nirvana: A Daughter's Life with Autism**]. There are hundreds of such stories, many of them verified.

All of these people were profoundly autistic, with poor or nonexistent speech and subnormal IQs. Such people do exist, but are exceedingly rare even within the autistic population. Less than 10% of autistics have so-called 'talented savant' skills (calculating, memory, musical or artistic ability far in excess of their generally retarded IQs), and an even smaller percentage have 'prodigious savant' skills (abilities far superior to that of the general population). It is believed that most savants and documented prodigies are autistic to some degree. Conversely, psychiatric assessments of autistic children have repeatedly confirmed the existence of uneven 'islands of ability' in perceptual, musical, or calculating skills amid relatively retarded cognitive or verbal performance. What accounts for this paradox? Is there some association between extreme disability and islands of ability or pure genius? In this chapter we will explore this issue in terms of imagination and dreams, memory, brain development, and the very concept of 'intelligence'.

Metaphors

In chapter 3, we saw that our perception of the world is not direct. Outside information is translated into electrical and chemical code and filtered through specialized modules in our brains until a useful and meaningful interpretation is made. The interpretation need not be perfect; indeed, it cannot be perfect. It is an approximation. Much of the detail, either at the initial level of perception or at 'higher' levels of selective attention, is left out. Some information is later 'filled-in'. But the final product is accurate enough to work with. Natural Selection has made sure of that. We don't think of our perceptions of the things in this world as being divorced from underlying reality. We take it for granted that the experienced world *is* the real world. Only on those occasions when we become consciously aware of a discrepancy between the two, such as after dreams or when reading German philosophy, do we make the distinction between reality and our perception of it. It is a distinction that, I think, is central to understanding the nature of the savant phenomenon.

The linguistic philosopher Ray Jackendorf has pointed out an interesting commonality in rather ordinary sentences like these:

The messenger went from Paris to Istanbul.
The inheritance finally went to Fred.
The light went from green to red.
The meeting went from 3:00 to 4:00.

All the examples are characterized by a metaphor of location: something moved. Yet only the first sentence concerns an actual location: the messenger physically moves in space. The other three statements deal with abstractions of possession, changes of state, and changes in time. In these examples there is no object that actually moves from location A to location B. Yet we speak (metaphorically) as if they do. The very fact that statements like these sound so natural to us even though they are not true may mean that there is a natural and possibly innate tendency for humans to think in abstractions using metaphors for physical actions, such as movement and force. Evidence for this comes from studies of children, who were found to spontaneously coin their own metaphors for desire, intention, and belief using the language of physical action.

Here are some actual examples from preschool children collected by the psychologist Melissa Bowerman (from Steven Pinker's **How the Mind Works**, pages 356-357):

You put me just bread and butter.
Mother takes ball away from boy and puts it to girl.

I'm taking these cracks bigger [while shelling a peanut].
I putted part of the sleeve blue so I crossed it out with red [while coloring].

Can I have any reading behind the dinner?
Today we'll be packing because tomorrow there won't be enough space to pack.
Friday is covering Saturday and Sunday so I can't have Saturday and Sunday if I don't go through Friday.

My dolly is scrunched from someone…but not from me.
They had to stop from a red light.

In addition to these, linguists have collected thousands of similar metaphors from every language ever studied. Such metaphorical representations of intentions are 'built upon' ideas about actions. This makes quite a lot of sense in terms of evolutionary psychology. Animals (yes, humans included) must constantly evaluate and negotiate their physical surroundings by calculating distance (to the nearest waterhole, fertile female, hiding place, etc) and force (to crack a nut, break a branch, push the bad guy off a cliff, etc). We know that there are discrete brain regions designed to accomplish these things (the dorsal visual system and the premotor cortex, respectively). We also know that natural selection does not create structures and modules from scratch. Body parts used for one purpose (fins for swimming, jaw bones for chewing) become co-opted for other functions in descendent species (limbs for crawling, middle ear bones for hearing). As Sir Francis Crick once supposedly remarked, 'God is a hacker'. An analogous process may have occurred in our ancestral primate brain—giving rise to a 'metaphor representation module' for understanding social action (Theory of Mind) from a vestigial 'action perception module' for understanding physical causality. While it is still very important for humans to understand and manipulate inanimate objects and calculate projectile trajectories, it is arguably even more important to understand and manipulate other people and calculate social decisions, because we are essentially evolved social animals. We constantly view virtually all aspects of our social life through the filter of intentional metaphor. If we could somehow remove those filters, the world would seem a truly strange and disjointed place; a place without free will, intentions, beliefs, or moral responsibilities; a world of funny looking organic automatons aimlessly wandering through a landscape of physical causality.

What is the nature of this innate metaphor representation system, if such a thing exists, and how is it related to TOM? The psychologist Alan Leslie has proposed the idea of a 'metarepresentational decoupler'. This is a dedicated mechanism or module in the brain that transforms representational knowledge of the physical world into an intentional proposition an agent has about an object. For instance, a mother holding a banana to her ear can be 'decoupled' into the metarepresentation, 'the mother pretends the banana is a telephone.' The agent (the mother) possesses an attitude (pretends-true) towards the thought content (is a telephone) referring to another object (banana). This mental grammar is highly versatile at representing all sorts of intentional stances that people may take: [a] [believes/does not believe/desires/does not desire etc] that [x] is [true/false y etc].

It is flexible enough to be used in recursive loops: 'I believe that she believes that he believed that they would have thought that it was false...'

The metarepresentational decoupler allows one to represent another mind's representation of something even if one knows that the second representation is wrong: 'Ptolemy *thought that* the sun and planets revolved around the earth.' This is the essence of 'first order' TOM. This can be elaborated into second or higher order TOM: 'Copernicus *believed that* Ptolemy thought that the sun and planets revolved around the earth.' Decoupling of belief states from reality lies at the core of social and cultural understanding. Without it, one could not appreciate sarcasm, satire, irony, wit; one would read novels, but miss the stuff 'between the lines'; listen to jokes, but miss the punchline. The person without a decoupler would lack imagination and creativity.

Not all metarepresentation involves TOM tasks. Abstraction can apply to physical and mathematical concepts as well. The ability of some higher functioning autistic individuals to successfully complete nonsocial abstraction tasks, such as higher mathematics, while failing TOM tests indicates that perhaps TOM is a special module at least partially distinct from, and perhaps 'embedded in' general metarepresentational ability.

Dreams

TOM and mathematical reasoning are just two examples of cognitive decoupling. A third is dream imagery. Let's now briefly examine the nature of dream sleep as a model for understanding the costs and benefits of the decoupling phenomena in general and in autism in particular.

The average adult spends six to eight hours of each day asleep. This consists of four to six complete cycles of REM (Rapid Eye Movement) sleep alternating with nonREM sleep. REM, which typically makes up 20% to 25% of adult human sleep, is associated with vivid dreaming. During REM sleep itself, there are short bursts of brain activity originating in part of the brainstem (the Pons), traveling to a relay station in the thalamus (the lateral geniculate nucleus), and terminating in the visual (occipital) cortex. These so-called PGO waves produce rapid eye movements, middle ear muscle contractions, and autonomic stimulation (changes in blood pressure, pulse, respiration, stomach secretion, etc.) in addition to vivid dreams.

Sleep experiment volunteers who were awoken in the middle of PGO bursts often report being in the midst of an intensely vivid and bizarre dream image of a hallucinatory quality, often unrelated to the 'background' dream they were having before. In contrast, when they were awoken in the so called tonic (non PGO) phase of REM, the subjects report being in the middle of a more coherent, but somewhat less visually vivid, dream. According to one hypothesis, the PGO spike acts to 'reset' the cortex with an image (via the brainstem), after which the cortex attempts to construct a coherent story or plot around it. The subcortical PGO component provides the underlying raw materials, which the cortical regions, especially the frontal cortex, clothe in interpretation.

According to the anthropologist Donald Symons, the content of dreams is largely visual and vestibular (the sense of movement such as floating through space), providing the illusions and hallucinations of sight and movement, but largely devoid of sound (except for speech), kinesthesis (movement of one's joints), temperature, touch, smell, and pain sensations. He believes that this curious asymmetry of dream experiences has evolved because, in his words, 'it is adaptive for individuals to be continuously alert and responsive to external stimuli (such as the sound and odor of an approaching predator or the cry of an infant), even during sleep. Natural selection thus has disfavored the occurrence during sleep of hallucinations that compromise external vigilance.' Symons calls this idea the 'vigilance hypothesis'.

During sleep, especially of the REM variety, our eyes are closed and we are essentially paralyzed (except in some abnormal conditions such as sleepwalking), presumably to protect the eyes and conserve energy. This allows for the vivid dreaming of images and of movement without the danger of acting out the dream. Moreover, the senses of touch, smell, temperature, pain and sound are not compromised, thus allowing us to monitor the environment subconsciously. Hallucinations in these sensory modes would be maladaptive, as the dreamer would confound dream with reality with possibly deadly consequences. On the other hand, the fact that we conjure elaborate scenarios that we 'believe in' while dreaming despite their lacking some key dimensions of perceptual authenticity perhaps reflects an innate drive for coherence. To quote Symons:

'…human beings, awake or asleep, are adept at conjuring engrossing scenarios from impoverished input (or in the case of daydreams, from no input at all) while ignoring the absence of sensory experiences that necessarily would be present if the scenarios were real. We quickly become engrossed in a movie's plot, for example, and we are captivated no less if the movie is in black and white, or is silent, or is a cartoon. We ignore the absence of color or sound, or the fact that we are watching animated line drawings, just as we normally ignore the fact that movies omit plot-appropriate sensations of odor, taste, touch, warmth, cold, and pain.'

[The stuff that dreams aren't made of: Why wake-state and dream-state sensory experiences differ. Cognition 47, 181-217 (1993)]

If we believe Symon's vigilance hypothesis, we accept that there are costs as well as benefits to dream sleep, and that nature has prevented us from making dreams too similar to reality. But in nature, no expensive habit goes unpunished unless it confers some advantage for the individual who possesses it. What, then, is the adaptive value of dream sleep? This is not entirely clear. Sir Francis Crick, for one, has proposed that dream sleep allows for the strengthening of neural connections that underlie cognitive associations made during wakefulness. Indeed, there is evidence that REM sleep is associated with the consolidation of memories of events and thoughts experienced while awake (although there are case reports of patients who for some reason have markedly reduced or nonexistent REM sleep but don't suffer from amnesia). On the other hand, healthy infants (who presumably experience fewer 'things to remember' during the day than older children and adults) have been found to spend up to 50% of their sleep in REM, and premature infants (who have no 'daytime memories' at all to consolidate) spend up to 100% of their sleep in REM. This suggests that dream sleep may also be important for the guidance of proper neural connectivity in the developing human brain.

Dreams are abstractions that carry costs and benefits. Natural Selection has compromised between the 'believability' of the dream world and the need to maintain vigilance of the wake world. Similarly, intuitive psychology is abstracted from the reality of intuitive physics. This too has costs and benefits. Individuals live and function at a certain personally optimal level of 'waking abstraction', which fluctuates depending on the task at hand. This level of metarepresentation has a population baseline that varies from individual to individual. Some people are good 'folk psychologists' who are naturally adept at social interaction. They excel as lawyers, writers, and actors. Others are better 'folk physicists' who gravitate towards carpentry, physical science, and accounting. An obvious difference between dream abstraction and waking abstraction is that we will all wake up (hopefully) after each night of sleep, but the baseline level of waking abstraction is, at least after childhood, generally set for life. This aside, we can imagine both phenomena as suspensions of belief above reality.

Let's continue this analogy. If we assume that autistic people are defective in abstraction, resulting in limited imagination and creativity, could we also expect autistics to have less imaginative dreams? My own feeling is that the vividness of hallucinatory episodes (the PGO phase) would not be any less intense in autistic dreams because this is a subcortical, and possibly more universal, phenomenon. However, the subsequent tonic dream phase, reflecting the more conscious cortical contribution, may be less creative. Devising an experiment to explore this fascinating question would be daunting because most autistic people are poor communicators. However, it would be consistent with a possible mechanism for autistic savant/prodigy skills: **subcortical facilitation at the expense of cortical function**.

Memory

Many autistic savant skills involve unusual feats of memory. Perhaps behind the savant phenomenon lies abnormal memory processing. Observers of autism have long noted the impersonal, automaton-like nature of savant memory. In his book, **Extraordinary People: Understanding the Savant Syndrome**, Darold Treffert writes, 'It is interesting that the prodigious, automatic memory of the savant is devoid of emotion. It stores trivia without ranking for importance and stores music without feeling the music—automatically, mechanically and literally. It stores with incredible accuracy and depth but is narrow and limited in its scope. It clearly differs from memory in the rest of us.' Many savants seem to reveal their particular skills spontaneously without prior training or practice. Importantly, they are rarely able to give any insight into their abilities when queried, sometimes simply saying that the talents 'come from God'.

This is very different from 'normal' memory. Skills are acquired with conscious, often painstaking effort. Without practice, performance is rarely perfect. We recognize that our abilities to drive a car, balance a checkbook, or speak a foreign language are not inborn or spontaneously generated skills. They are mastered through learning. Once learned, they become 'part of us'. There is an intimate relationship between our particular skills, knowledge, and memories and our sense of personal identity. But in autistic savants, the special ability to calculate square roots, memorize pi, play piano, or paint a perfect landscape are strangely dissociated from the rest of their personalities. There is something alien and otherworldly about these remarkable skills.

We can gain some insight by taking a brief stroll down memory lane. Memories can be categorized as episodic, semantic, and procedural. Episodic memories, also called autobiographical memories, are the recollection of specific events in one's biography, such as the day in the fifth grade with which I began this chapter. Semantic memories refer to general knowledge of things in the world, as opposed to a specific personal episode. It is the knowledge that Kuala Lumpur is the capital of Malaysia, that the parathyroid glands are part of the endocrine system, or that schadenfreude is the German word referring to taking pleasure in another's misfortune. Procedural memory is used to carry out an action, such as one's backhand, backstroke, or fox trot.

The three types of memories can be further subdivided into short (seconds), intermediate (minutes to days), and long term (days to years) memory. Short-term memories, such as remembering a new phone number long enough to write it down, are believed to involve transient reverberations of cortical activity after the initial perception. They may be associated with activity in the prefrontal cortex, especially if the memory involves a motor response ('working memory'). Intermediate memory is acquired, or 'encoded', via the medial temporal cortex and a very important subcortical organ in this region, the hippocampus.

The hippocampus receives sensory information from the visual, auditory, and somatosensory association areas and stores the input using a specific spatial and temporal code. The contour of a face, the scent of a perfume, the layout of a room are linked to each other and to the coordinates of the viewer's spatial location at a particular time. This code is stored for a time in the hippocampus and slowly channeled into the adjacent medial temporal and more distant cortical areas, where it becomes long-term memory. The hippocampus is the transfer station from intermediate to long-term storage in the cortex for most episodic and semantic (but not procedural) memories. The amygdala plays an important supporting role in the encoding of emotionally salient memories. This is why we tend to remember a particularly poignant or emotional episode with so much more clarity than something run-of-the-mill.

Procedural memories, in contrast, are not believed to require the hippocampus for encoding. Rather, they are stored directly in the brain regions that first mediated the activity. For example, riding a bicycle involves activity in the visual and motor cortex, in addition to the subcortical regions responsible for balance and motor planning, the cerebellum and basal ganglia. With practice, these tasks become increasingly automatic. Eventually we learn to ride a bicycle 'with our eyes closed'. Consequently, the brain regions active in mediating an 'overlearned' task are reduced to subcortical regions, with little activity in the cortex.

The retrieval of memories also appears to differ between procedural, semantic and episodic memories and between intermediate versus long term memory. Patients who have suffered bilateral hippocampal damage experience anterograde amnesia (the inability to make new memories) both for specific events and for general facts. However, they can still learn new mechanical skills because procedural memory does not require the hippocampus. Additionally, amnesiacs tend to suffer retrograde amnesia (inability to access memories made before the injury) going back several days, weeks, or sometimes years depending on the extent and severity of damage, again for semantic and episodic memories, but not for procedural skills. (**figure 1**) It is believed that the hippocampus eventually transfers the 'memory traces' into the cortex over months to years, thus remote and childhood memories are paradoxically the 'last to go' following hippocampal amnesia or dementia. In summary, the hippocampus is required for intermediate and long-term encoding of semantic and episodic events (and for intermediate recall of these events), but is not involved in long-term memory retrieval or in the processing of procedural memories.

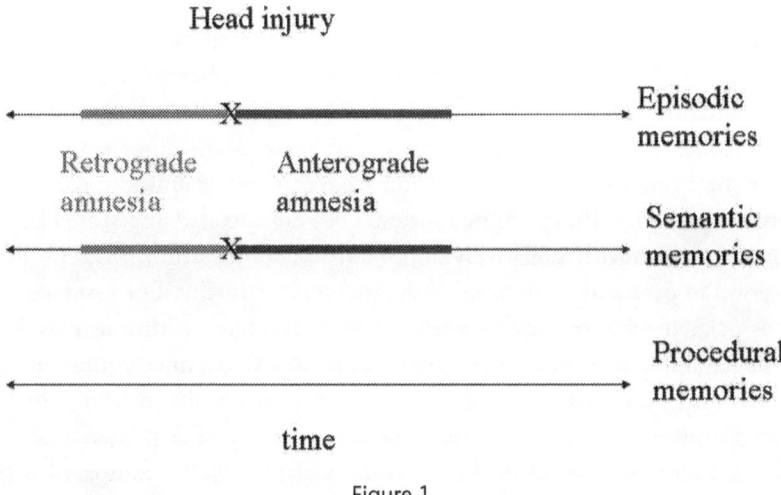

Figure 1

This means that there are separate systems for the encoding and retrieval of automatic, mechanical skills, on one hand, versus conscious knowledge about the world and oneself, on the other. The former relies preferentially on subcortical organs (the basal ganglia and cerebellum), while the latter requires the cortex and hippocampus. Unlike normal people, savants may use the subcortical pathway to store skills such as piano playing, drawing, and even lightning calculations.

To complicate matters, the two types of cortical memories appear to be stored in different brain areas. Imaging studies using fMRI have demonstrated that the retrieval of **episodic memories** is associated with stereotypical activity in certain parts of the frontal and parietal lobes. In contrast, **semantic recall** tended to be associated with activation in parts of the lateral temporal cortex. Some call this area of the temporal lobe 'semantic space' because knowledge of things in the world seems to be stored there. Amazingly enough, this abstract semantic space is actually anatomically organized into categories. Knowledge of living things (such as 'elephant'), man-made artifacts (such as 'hammer'), and faces appear to be segregated to specific areas of the temporal cortex.

The frontal cortex appears to imbue memories with a sense of personal intimacy. Patients with damage to the frontal cortex, due to dementia or stroke, sometimes describe having strange recollections of events which they don't think they have actually experienced.

I propose a 'memory hierarchy' (**figure 2**). Episodic/autobiographical knowledge, being the most personally salient or 'conscious' type of memory, is at the top. It is encoded in the hippocampus/medial temporal lobe, stored in various parts of the neocortex, and requires frontal/parietal activity for retrieval. Below this is semantic/'rote' knowledge, which is less personal, but still allows conscious access. Semantic memories are also encoded by the hippocampus, stored largely in the nearby temporal cortex (semantic space), and are more easily retrieved without the frontal/parietal contribution characteristic of episodic recall. At the bottom are procedural/implicit memories. They are encoded and stored largely in subcortical regions with a relatively minor cortical contribution. Savant skills may correspond to overactivity in parts of the memory hierarchy. For instance, savant skills associated with the less severely autistic Asperger syndrome may involve superior semantic memory. More severely autistic savants, once commonly called 'idiot savants', may have superior procedural memory skills instead. This could explain the more self-conscious, self-reflective capacity of Asperger savants who are perhaps more deliberate in the cultivation of their skills, compared with the more automatic skills of the autistic 'idiot' savant.

	Location	characteristic
Episodic Memory	frontal/parietal/temporal/ hippocampus	conscious/personal
Semantic Memory	temporal/ hippocampus	conscious/impersonal
Procedural Memory	basal ganglia/ cerebellum	subconscious

Figure 2

Hidden access

Superior memory systems can explain many aspects of the savant phenomenon, but not all of them. Certain skills, such as musical ability, perceptual sense, and some forms of arithmetic calculation do not involve feats of extraordinary memory. What is the basis for these skills? According to the psychologists Allan Snyder and John Mitchell, some savant abilities appear to result not from superior memory, but from 'privileged access to lower levels of information.'

Artistic autistic savants seem to have an innate appreciation for visual perspective. This allows them to automatically transpose the three-dimensional world they see directly into the two dimensions of the paper. The lightning calculator savants appear to actually perceptualize large numbers as discrete quantities in their mind's eye, rather than as qualitative approximations, as children do, or as digital abstractions, as adults do. Thus they are able to directly manipulate integers for rapid calculations. The musical savants may have an innate talent for portioning a musical piece into elements of tonality and scale. They can 'feel' and play the music without first having to consciously translate it into a conceptual abstraction, as most normal musicians do. Other savants may have an innate 'sense of time' enabling them to reflexively tell the exact time when awoken in the middle of the night, for example.

Neurotypicals are unable to represent or manipulate number, perspective, pitch, and time in such a direct and automatic manner. As children, we are painstakingly taught how to draw in perspective (and most adults are still lousy at it), multiply and divide numbers in longhand, read music, and tell time on a clock face. These skills don't come naturally. They are acquired through repeated conscious processing through cortical circuits. Perhaps in the autistic savant subcortical processing dominates. Just as children learn to ride bicycles and speak foreign languages more quickly than adults can, autistic savants learn to calculate dates and read music more easily than a normal person can.

Neither our perceptions nor our memories are direct reflections of the real world. Like the contents of our dreams and the theory of our minds, memories are creative interpretations and reinterpretations of sensory experience. The English psychologist Sir Frederic Bartlett writes in his classic book **Remembering**:

Remembering is not the re-excitation of innumerable fixed, lifeless and fragmentary traces. It is an imaginative reconstruction, or construction, built out of the relation of our attitude towards a whole active mass of organized past reactions or experience, and to a little outstanding detail which commonly appears in image or in language form. It is thus hardly ever really exact, even in the most rudimentary cases of rote recapitulation, and it is not at all important that it should be so.

I believe that memory, like perception, is a continuously creative process with multiple neural traces dynamically etched out in the hippocampus and cortex with each recollection. We are actually remembering a memory of a memory of a memory…editing it slightly each time we 'relive' the original experience. It is intriguing to think that if brain damage were to somehow affect the natural ability to creatively conjure abstractions, perhaps a privileged access to a lower, hitherto hidden, level of memory and perception would be gained.

The notion that autistic savants have privileged access naturally begs some questions. Do each of us have the potential for such special access that is either lost or never realized? Can this access somehow be regained through training? The obvious answers would be negative. It is true that savants do not appear to learn or improve much upon their skills with practice. But even well trained mathematicians cannot usually match the lightning fast savants in sheer speed of mental calculation. It is not at all clear if such skills are something unique to savants or if they exist in all of us, but are somehow suppressed in the process of cognitive development.

There is evidence for the existence of enhanced perceptual, memory, or motor abilities (called 'paradoxical functional facilitation' or PFF) following brain damage. For example, some patients with epilepsy who underwent split-brain surgeries were subsequently found to perform better at simultaneous left and right handed tasks (try tapping your head while rubbing your belly) than normal controls. There are even reports that some subjects of near drowning or head trauma recover with *enhanced* memory or perceptual abilities. The nature and origin of PFF is complex and is likely to involve many possible mechanisms. We will examine some of these mechanisms in an effort to understand the concept of developmental plasticity in autism.

Plasticity

The mental construction and imposition of meaning and coherence on the world is a natural human tendency. Meaning and coherence are by definition imaginary constructions. Memories, beliefs, desires, and even perceptions are products of the creative mind: metaphors built upon the real world of quantum physics, probability, and thermodynamics. There are mechanisms in the brain that build the illusions that we experience and call 'reality'. Perhaps there are some individuals who, through inborn or developmental defects, have a more privileged view of the 'lower levels of reality'. The price they pay is a less creative and coherent worldview.

The natural drive to pull disparate information together in terms of background context seems to occur at the level of neural connections. For example the receptive fields of the primary visual cortex are not passive processors of visual input. Rather, they dynamically change in size depending on the context of surrounding sensory input. Normally, neurons are activated by appropriate stimuli in their receptive field (such as a vertical line or a blinking dot) and inhibited by stimuli outside it. But an appropriate stimulus, such as a long straight edge that activates several neighboring cells, will produce one large receptive field. These dynamic changes, as described by the neuroscientist Charles Gilbert and colleagues are known as 'cortical plasticity'.

Cortical plasticity explains, among other things, the phenomenon of perceptual 'filling-in', in which a portion of the visual field that is inactive (for example, our natural 'blindspot') or damaged is still able to 'see'. This happens when the neighboring RFs increase in size to 'fill-in' the gap. Peripheral perception flanking an 'attended' central location provides the context needed to extrapolate the gist of the whole (object, situation, experience, memory, meaning, etc).

Cortical plasticity comes in several flavors. It may occur over the short term (seconds or less) via the unmasking of inhibitory neural connections. Gilbert found that monkeys that had a discrete area of their visual fields occluded (temporarily) experienced an immediate growth in the size of the flanking receptive fields, as measured by electrodes implanted within the visual cortex. When the occluding stimulus was removed, the receptive fields immediately shrunk to their normal size. This correlates well with similar (but less invasive) experiments conducted on human volunteers.

Alternatively, plasticity may occur over many weeks or months, likely involving the growth of new neural connections. Dramatic examples of long-term plasticity include V. S. Ramachandran's discovery of phantom limb representations in the trunk and face of amputees and M. Merzenich's work on reorganization of the body representation maps in the brains of monkeys whose fingers were experimentally amputated or sutured together. In these cases, abnormal or non-existent input from the outside produced a permanent change in neural connections within the brain such that novel or even enhanced perceptual abilities arose.

Ramachandran calls this the 'remapping hypothesis'. The evidence suggests that both long and short-term remapping occur through the growth of new synapses and the use of existing connections that were previously inhibited.

Abnormal cortical plasticity of the long-term variety may be at the core of autistic savant skills. Normal development is achieved via a complex interaction of genes, the prenatal environment, and lifetime experience. The end product of nature and nurture is a plastic mind continuously molded to function in its particular physical and social environment. Without proper instructions, perhaps due to defects in the growth of neural connections or in the appearance of normal inhibitory modulation during early development, the mind may remain fixed in a more primitive 'cognitive style'. It is a style with a characteristic blend of strengths and weaknesses. It allows for a paradoxical facilitation of subconscious perception, calculation, and memory at the expense of conscious, reflective control. This is the kind of mind we find in the autistic savant.

Intelligence

I started this chapter with memories of a precocious childhood. The precocity has been lost. I no longer collect mental lists of kings and capitals, mountains and islands, space probes and tennis stats. Does this make me less intelligent? People don't go around calling me 'little professor' anymore. But neither do they call me 'nerd' or 'geek'. Perhaps I have exchanged some intellectual precocity for social maturity?

Intelligence is a difficult concept to define. We can categorize and compartmentalize it into things like 'verbal IQ' and 'performance IQ' or 'social intelligence' versus 'mechanical intelligence'. These definitions are not very rigorous. One measure of a basic component of intelligence is 'inspection time'. Inspection time (IT) is the minimum time an experimental subject needs to be exposed to two simple visual stimuli (such as a pair of parallel lines of unequal lengths briefly flashed on a computer screen) in order to make a correct discrimination. IT is not to be confused with 'reaction time'. The time limitation is on the duration of the stimulus (the period of time the image is on the screen), rather than the time it takes to respond. The subject gets as much time as needed to make the decision. It is believed that IT gives a rather good approximation of cognitive processing capacity. IT corresponds well to other measures of intelligence in the testing of, for example, mentally retarded children. However, Uta Frith and colleagues have found that autistic children with measured IQs one standard deviation below average performed as well as normals in a test of inspection time. They conclude that autism is associated with a deficit in a component of intelligence which has little to do with basic information processing. The missing component may be related to social insight. It may be the abstraction ability that enables creativity, imagination, and a theory of mind.

Earlier we saw that there is a population gradient in abstract representational ability or 'imagination' with most people falling somewhere in the middle of the bell shaped distribution curve. (**figure 3**) At one extreme of this curve are those people with a profound deficit in social awareness, as is characteristic of the autistic spectrum disorders. At the other extreme are those with 'too much' imagination. These individuals are prone to attaching meaning to and seeing significance in things and situations in sometimes maladaptive ways. They suffer delusions and hallucinations. Schizophrenia and drug-induced psychotic states fall into this part of the curve. Between these extremes are the majority of 'neurotypical' people. Some are folk physicists, others are folk psychologists, but all are socially functional to some extent.

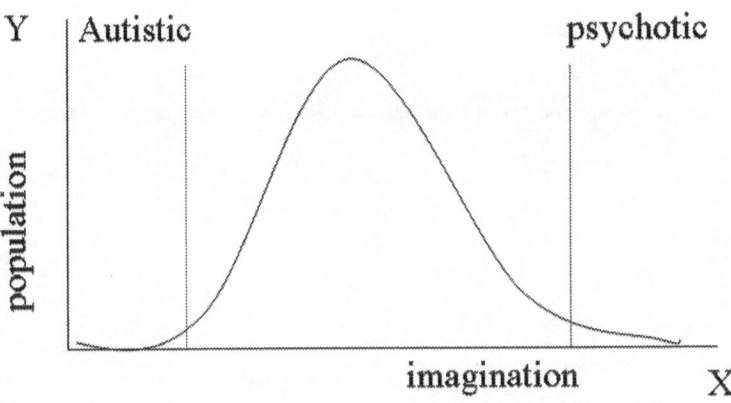

Figure 3

It would be simplistic and misleading to suggest that schizophrenia is somehow the 'opposite' of autism. But I believe that this is a useful way of thinking about the nature of metarepresentational dysfunction in two very different disorders of thought. An analogy would be the idea that schizophrenia is the neurochemical opposite of Parkinson's disease. The former is a disease of thought, involving overactivity in the dopamine transmission in the frontal cortex, while the latter is largely a disease of movement, involving underactivity of dopamine transmission in the basal ganglia. The manifestations of these two disorders are very different and it would be meaningless to call them 'opposites'. Yet both are related to levels of a particular neurotransmitter. Similarly, schizophrenia and autism can be regarded as cognitive opposites. Intriguingly, neurotransmitters may underlie this dichotomy as well. As we shall see in the next chapter, serotonin (a chemical related to dopamine) may have a role in the pathogenesis of autism.

Basic intelligence, as defined by ITs or other measures of cognitive processing power can also be plotted on a bell curve. (**figure 4**) At the extremes of the distribution are those individuals with unusually fast or efficient versus those with retarded information processing. It is important to point out that this distribution reflects potential rather than functional intelligence. In the real world, the relative cognitive potential that we all have is combined with the imaginative ability demonstrated in the previous curve. We can combine the two curves into a composite 3-D model. (**figure 5**)

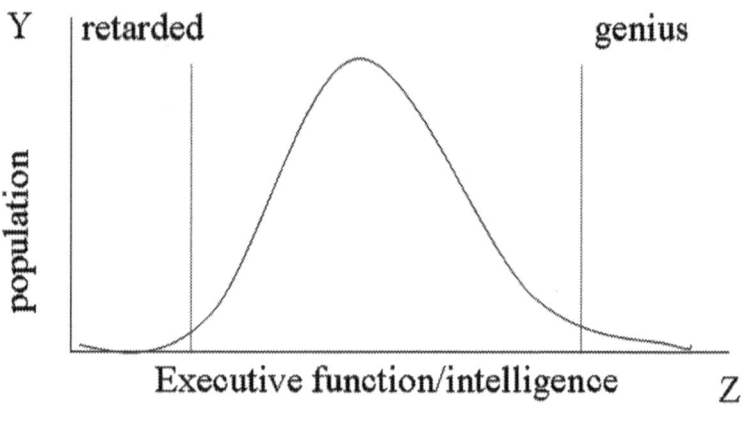

Figure 4

The model plots imagination on the 'x' axis, information processing on the 'z' axis, and population on the 'y' axis. The shaded regions represent individuals in the suboptimal functioning range, either due to retarded intellectual processing (low z), defective imagination (low x), or hyperactive imagination/delusional thought (high x). Psychotic and schizophrenic individuals would be found in 'zone B' and 'psychotic geniuses' are labeled 'B+'. People with autistic spectrum disorders are in 'zone A', and a subset of those with high processing capacity, the autistic savants, are labeled 'A+'.

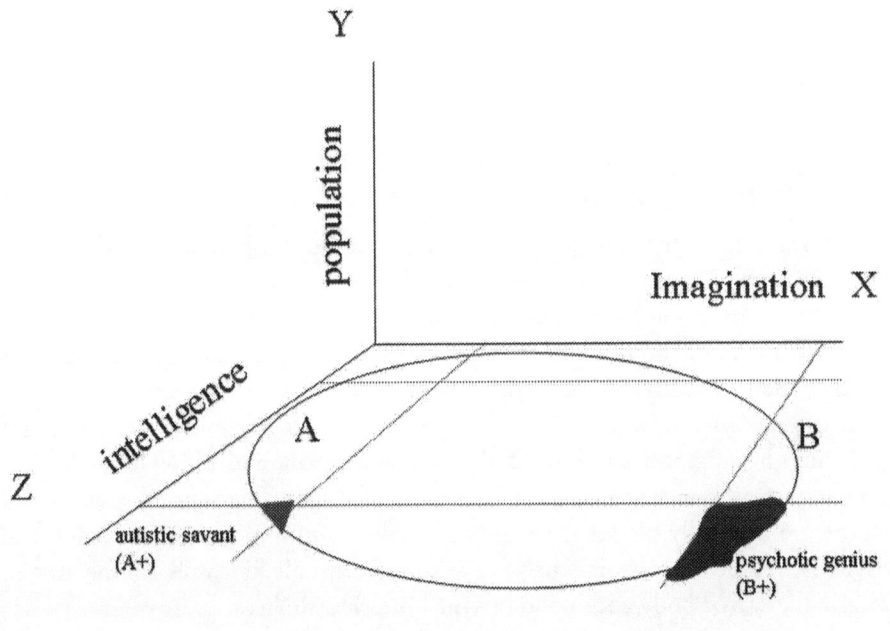

Figure 5

7

Sex

I'm standing at the pew lip-synching the Hymn of St. Francis. It is Good Friday, 1981, and I haven't been a student at Holy Innocents School long enough to learn the words to the hymns and prayers, or even the sign of the cross (is it right shoulder, then left shoulder, or the other way around?) like all the Catholic students. Father O'Reilly is struggling through one of his last services. He will be dead of emphysema in another year. Between spasms of violent coughing and wheezing, he implores us to avoid sin and temptation, and to avoid the evil of smoking, so as to enter the Kingdom of God. The agonizing sermon goes on and on, before he finally blesses the wine and Holy Communion, the cue that I will have to kneel down conspicuously while all the Catholic kids walk up the aisle to receive the sacred body and blood of Our Lord, Jesus Christ.

I sit alone, the single atheist in the eighth grade. My parents transferred me here a month ago in response to the daily harassment I suffered at the hands of my peers at the local public school. But this is worse. Now I feel totally alienated, separated not only by psychology, but also by religion.

'Our Father who art in Heaven…' I feel disoriented and lightheaded. '…hallowed be thy name. Thy Kingdom come, thy Will be done on earth as it is in Heaven…' I can visualize the heavy wooden canopy above the altarpiece collapsing onto the frail body of Father O'Reilly, killing him instantly and liberating me from this misery. '…for thine is the power and the glory of God as it was, is, and ever shall be, world without end, amen…'

◆　　◆　　◆

'…nobis peccatoribus vitam aeternam…,' mumbles the silver-haired old Master at the High Table. Everyone sits down. A lanky young man in a rugby shirt nonchalantly walks across my table, deftly stepping between tureens of gravy and plates of mushy peas. Boisterous laughter erupts from a group of arrogant looking young men with upper-crust British accents. It is the start of Michaelmas term at University College, Oxford, 1990.

What am I doing here? I ask myself, spending a year and a big chunk of my parents' money to study what…British imperial history and political philosophy? Once held goals become fuzzy with doubt. As if it weren't bad enough trying to fit in at MIT, where most people were social misfits anyway, now I'll have to deal with the elite of the English class system. How do I begin to pretend to be normal?

'Hey, Ian and I are going to King's Arms tonight, you comin'?' The gruff taunting voice belongs to Andrew, my housemate and quintessential ugly American. He is as obscene as he is self confident, and displays a rakish charm that the English girls inexplicably find fascinating. A wave of anxious nausea washes over me, but I put on a game face.

'I have a philosophy tutorial to prepare for tomorrow, but…'

'But nothing, you loser! You think you're gonna' get laid talking like that?'

Andrew is becoming a malignant influence. The previous weekend, he and I biked from Oxford to Blenheim Palace, a distance of ten miles, stopping at five pubs along the way. After a hazy tour, we biked back stopping to repeat the pints at each of the pubs. I was nearly killed on the way back.

'Allright, I'll go.'

He slaps me a high five and then looks over at a young woman walking towards our table. 'Hey guys, speaking of pussy, here's Sarah. Watch this.'

Sarah was the very English and very attractive coxswain of the University College women's boat. She had met Andrew at the orientation party the first week of term, spent the following Saturday night in our flat, and was now in the process of breaking up with her Scottish boyfriend.

'Sarah, sweetheart, what's the matter? You don't want Nigel seeing you with the Yanks? Come on, I want to introduce you to some great Americans here. That's Sandy, our grad student from Berkeley. He's training for the Olympic team and gunning for the all-Oxford blue boat. That's Matt. He's a Marshall Scholar at Christchurch. Oh, and I almost forgot, that's my housemate Henry, the MIT geek.'

Andrew gives me a brief look before turning his attention to Sarah's face, laying a hand on her slim waist and pulling her down to his lap. I am speechless with a toxic mix of humiliation, resentment, and envy.

'I was just telling Sandy here how good English women are in bed. He doesn't believe me, yet.' Andrew raises his glass with a complacent smirk. Sarah promptly blushes beet red, stares at the ground, and manages a bashful smile. Everyone laughs. I have an out of body experience.

Part I: Evolution

Some people, like the acerbic New York Times columnist Maureen Dowd, believe that there is a battle between the sexes for world domination. If so, which is likely to come out on top? Some might argue that males are the stronger sex based on the overwhelming evidence we see every day in the modern world. Men dominate political discourse and economic activity. Most of our laws are made (and broken) by men, enforced by men, and adjudicated in courts presided over by mostly male judges. Men make (and spend) most of the world's wealth. The wealthiest CEOs have grown fat by commanding the labor of less fortunate male muscle. From the expendable infantryman to the most decorated field marshal, men totally dominate military adventures. Men get to do almost all of the fighting and most of the dying. The greatest scientific and technological achievements in human history have been made largely by men. Even in the realm of the arts, traditionally considered a more 'feminine' domain, men have dominated. The best known and most prolific painters, sculptors, poets, composers, and rock stars have been men. Women may do most of the world's cooking, but the most acclaimed chefs in the world are men. Women may wear the most fancy clothes and shoes, but chances are they have been designed by men. It seems rather obvious that males are the stronger sex. How can we explain, if not justify, male domination in so many facets of human endeavor throughout recorded history? Perhaps their superior physical strength somehow translates into a more aggressive psychological drive for creativity?

I'm sure most people in our supposedly enlightened, 'politically correct' society will find the previous paragraph both sexist and morally insulting. While I don't think a perceived moral impropriety is a reason to overthrow any theory, especially one that has ample evidence in its support, at a deeper level, I do believe that the idea of male superiority is seriously flawed. It is true that males dominate most of the institutions of society and in so doing have largely driven the course of human history through their wars, laws, trade, ideas, and inventions. But it is only a small fraction of men who have actively shaped these processes. Most women and men have been innocent bystanders. And more importantly, civilization itself is only 5000 years old at the oldest (Mesopotamia and Egypt) and far younger in most other parts of the world. Modern human beings, by contrast, have been around for over 100,000 years, and our earlier hominid ancestors date back several million years. For almost that entire period, there were no political parties, economic philosophies, stock markets, factories, organized wars, universities, or art galleries for males to dominate. The only rock groups were the ones outside the cave. I argue that for the vast majority of human history, men and women were, in terms of their utility and perhaps in social standing, probably quite equal in their different and complementary ways. Men hunted; women gathered. Both were equally important. The phenomenon of male domination may be a recent and perhaps temporary accident in the long course of human evolution. This is a question for cultural anthropologists. But there are other, more damaging attacks one can make on the notion of 'male superiority'.

Evolutionary psychologist David Buss, in his illuminating work on human sexual behavior, **The Evolution of Desire**, writes:

'Human mating mechanisms account for the puzzling finding that men die faster and earlier than women in all societies. Selection has been harder on men than on women in this respect. Men live shorter lives than women and die in greater numbers of more causes at every point in the life cycle. In America, for example, men die on average six to eight years earlier than women. Men are susceptible to more infections than women and die of a greater variety of diseases than women. Men have more accidents than women, including falls, accidental poisonings, drownings, firearm accidents, car crashes, fires, and explosions. Males suffer a 30 percent higher mortality rate from accidents during the first four years of life and a 400 percent higher mortality from accidents by the time they reach adulthood. Men are murdered nearly three times as often as women. Men die taking risks more often than women and commit suicide more often than women. The ages between sixteen and twenty-eight, when intrasexual competition reaches a strident pitch, seem especially bad for men. During those ages, men suffer a mortality rate nearly 200 percent higher than women.'

Researchers have long noted that the incidence of miscarriage, premature births, birth defects, infant mortality, and neurodevelopmental disorders are all significantly higher in males. Boys are more likely to suffer from mental retardation, dyslexia, stuttering, attention deficit disorder, Tourette's syndrome, and cerebral palsy than girls. One study cites that 'boys had a 20% higher risk for a low five-minute Apgar score and an 11% higher risk for being preterm. After the perinatal period, boys were found to have a 64% higher cumulative incidence of asthma, a 43% higher cumulative incidence of intellectual disability, a 22% higher incidence of mortality, and a higher...incidence of epilepsy and vision disorders.' [**Gissler, M., et al 'Boys have more health problems in childhood than girls: Follow-up of the 1987 Finnish birth cohort' Acta Paediatr 1999; 88(3): 310-314**]. Although male conceptions outnumber females 115 to 100, and 105 baby boys are born for every 100 baby girls, by the age of 100, women outnumber men eight to one. [**'The Weaker Sex' by Maggie Jones in The New York Times Magazine, March 16, 2003**] And finally, something more relevant to our topic: autism is up to four times more common in boys and Asperger syndrome may have a male/female ratio approaching a whopping 10:1 according to some studies. [**Baron-Cohen, S. (1999).'The Extreme Male-Brain Theory of Autism' in H. Tager-Flusberg (Ed.), Neurodevelopmental Disorders**].

So, which is the stronger sex? If males are so much more vulnerable than females as the studies seem to suggest, how have men come to dominate culture and society the way they have? I think the question may be improperly formulated. The real issue is that **males tend to be weaker than females** on many physiological indicators **as a group**, but that **some individual males are much more fit** physically and perhaps mentally than the average male or female, and that these males are the ones that tend to take dominant roles in modern human culture. When we see male domination of politics, economics, art, and science, what we are witnessing is the tip of the male iceberg, so to speak. The invention of agriculture, written language, centralized government, mechanized weaponry, and the resulting unequal distribution of accumulated excess wealth which characterizes modern human civilization, have combined to make modern culture increasingly more sensitive to control by small groups of strong individuals. These individuals have tended to be men.

The Weaker Sex?

Imagine a simple graph. On the X-axis, we plot some aspect of health or ability, such as cardiovascular condition or general intelligence. On the Y-axis, we plot the number of people. If we graph a large random population of women (for instance, 'all women who live in Manhattan'), we get a certain bell shaped curve (**figure 1**). Similarly, we can graph all the males in Manhattan to generate another bell curve. The curves will look similar, but if we superimpose the two graphs, chances are they will not fit exactly. The male bell will be slightly wider and flatter than the female bell. We could repeat this experiment for different populations, say men and women in Peoria, Illinois, or Oxford University, or the!Kung bushman tribe of Botswana, and consistently arrive at the same result:

the male distribution is always slightly wider. Of course, the actual shape of the curves will vary depending on the population used and the trait measured, but, in general, the pattern should hold. Why?

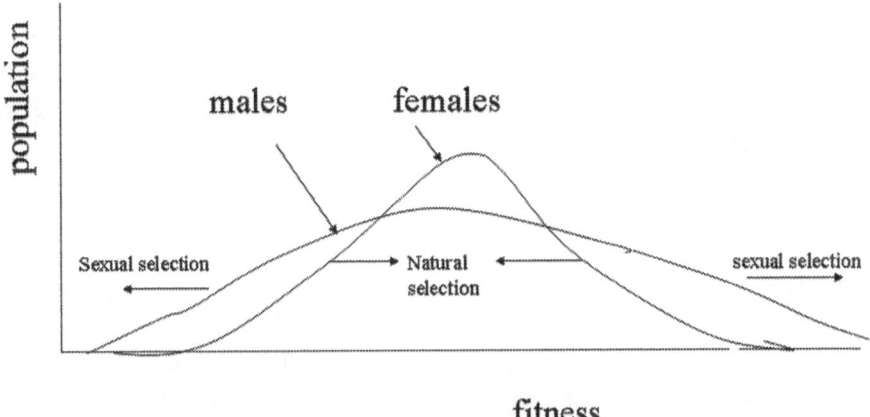

Figure 1

To see why this is so, we need to understand how evolution works. There are three basic concepts in Charles Darwin's evolutionary theory: **variation, heritability**, and **selection**. Variation means that individuals will have differences in certain traits; for example, Gregor Mendel found that garden peas can differ in color (yellow vs. green) and in texture (smooth vs. wrinkled). Those simple physical traits are determined by separate genes—one for color and another for texture. The color gene comes in different 'flavors' or alleles, one for yellow and another for green. Likewise, the texture gene comes in different alleles as well; one produces a smooth pea, another produces a wrinkled pea. This discovery of the so-called 'mendelian traits' paved the way for classical genetics, as we shall see later. Humans differ in various traits as well, such as height, eye color, hair texture, intelligence, and personality. Eye color is a simple mendelian trait that is determined by the alleles of a certain eye pigment gene. Traits such as intelligence and personality are much more complicated, involving the interaction of hundreds of genes with each other and with the environment. And of course all individuals are composites of many different traits. Thus the variability in a population can become very large indeed. Moreover, new variations are constantly being added to the gene pool through the natural process of mutation.

Heritability simply means that those traits coded by the DNA in one's genes can be passed down to one's offspring. Half of the offspring's genes come from the mother, half from the father (and roughly a quarter from each grandparent, an eighth from each great-grandparent, and so on).

Selection means that not all individuals will survive to reproduce and pass on their genes. This is the well-known 'survival of the fittest' concept. But there is another type of selection, also discovered by Darwin. This is sexual selection and we will need to examine it in more detail, for it may have been even more important in shaping the human mind.

Sexual selection refers to the process of reproduction as practiced by sexual organisms. In most sexual creatures, such as oak trees, orchids, octopi, and ourselves (but not bacteria, mushrooms, and certain types of insects and plants, which reproduce asexually by 'cloning' themselves), there are two sexes. Males produce sperm, females produce eggs. Males and females seek out and select each other based on certain preset criteria. Some of the criteria may appear arbitrary, such as the brightness of a peacock's tail or the color of one's hair. Others may reflect some obvious utility, such as the size of one's biceps or bank account. At any rate, one individual (usually female) tries to choose the best possible individual of the opposite sex with which to mate.

In most species, evolution occurs via both natural and sexual selection. Natural selection tends to decrease variation because individuals who possess traits that are ill adapted to the environment tend to survive less well and the genes coding for those traits are eventually eliminated from the gene pool. Animals with poor eyesight or weak legs or bad judgment may get eaten, starve, or fall off a cliff. None of these outcomes are conducive to having children. Only those individuals that are well adapted to survive to sexual maturity can pass on their genes.

Sexual selection, unlike natural selection, tends to increase variation through a process called 'fitness spreading'. Fitness spreading occurs when the fittest individuals select the fittest members of the opposite sex, slightly less fit males pair off with slightly less fit females, and so on down the line to the least fit members of both sexes. This is usually what happens in moderately monogamous species, such as humans. As long as the population is fairly monogamous, and not all the females are scooped up by a few powerful men, even the least fit males will still have a reasonable chance to mate and pass on their genes. Male variation will then increase through the accumulation of new mutations. In a highly polygamous population, where one male dominates a harem, we would expect to find somewhat less variation because most males will not add their genes to the pool.

Mars and Venus

Now let us discuss two possible reasons for the different extents of variation in males and females. The first is related to **natural selection**. Having to give birth to and raise offspring puts certain biological constraints on females. Women have biological costs imposed by the necessity of producing eggs, carrying the unborn fetus for nine months, and nourishing the infant for several more years. Males, in contrast, produce millions of expendable sperm, don't carry the child inside their body, and don't produce milk. A male's reproductive investments are much smaller and therefore his costs are proportionally less. Mutations are likely to be detrimental to the complex process of female reproduction. Therefore, natural selection has kept females from experimenting with the full range of variation; it has, in effect, narrowed the female bell curve. There is evidence to support this argument. Animal species in which the female lays eggs and has little or no role in the subsequent rearing of her offspring, including many types of fish and reptiles, show much less sexual variation in fitness. This is because the females suffer little more reproductive cost than the males. In these animals, being at the extremes is no better or worse for one sex than for the other.

The second explanation for why nature has chosen to take more chances with males has to do with choice. This is the argument from **sexual selection**. It starts out from where the natural selection hypothesis left off: the differential costs of reproduction. Males of most species invest little time and energy on the making and raising of offspring. They waste millions of sperm per ejaculate, each of which can potentially fertilize a single egg to produce a complete individual. By contrast, the female spends much of her precious resources on the production of a much more limited supply of eggs. A normal human female can expect to release around 400 mature eggs from her two ovaries during her reproductive life-time. Because single eggs are so much more expensive to make than single sperm, females have naturally evolved to be choosy about the source of the sperm they select to fertilize their eggs. The analogy can be extended beyond the germ cells: individual males, like sperm cells, are more 'expendable' than individual females.

An undiscriminating female is more likely to bear defective offspring than a choosy female. In addition, her female offspring might also inherit the tendency to be less choosy. The weaker and less choosy daughter of the less choosy mother and less fit father would in turn have a decreased chance of mating and having healthy offspring herself, or even of surviving at all. The unfortunate mother would be more likely to get stuck caring for defective offspring, wasting resources that could have been better spent on raising healthy offspring, who, in turn, could pass on their genes to the next generation.

Males, on the other hand, have relatively little incentive to care for their children because making sperm is so much cheaper and easier than making eggs, surviving pregnancy, and producing milk. It is more economical for a male to have as many offspring with as many females as possible, with the expectation that at least some of them will survive to reach adulthood, mate with more females, and propagate his genes.

Animal species (like humans) characterized by male competition for access to choosy females tend to have greater male variation. The variation goes beyond body types and facial features. It extends to physical and mental fitness. Males tend to outnumber females at the extremes of intelligence, health, and creativity, while females tend to outnumber males in the middle of the bell. Support for the sexual selection theory of increased male variation comes from the few cases in the animal kingdom where the 'traditional sex roles' are reversed. Among the Chicana birds of South America, for instance, females lay the eggs, but 'harems' of males incubate them. The larger females compete with one another for control of sexual access to the smaller males, and as a result, there is greater female variability.

Natural selection limits the width of the female fitness curve through biological constraints, while sexual selection widens the width of the male fitness curve through increased female choice (**figure 1**). Sexual selection, which tends to increase diversity through fitness spreading, and natural selection, which works to cut down on variations at the lower extreme of fitness, are the dual engines that power Darwinian evolution.

The fact that the best males are the most reproductively successful implies male competition for sexual access to females. This is why it pays for males to evolve extreme, potentially self-destructive physical and behavioral traits that sometimes lead to brilliant success, but more often to failure and death in the struggle for ever more life, sex, and offspring. To the few lucky winners, go the spoils of victory: greater sexual access to females and control of those material resources (wealth, status, power, prestige, etc.) that women find so attractive. The traditional feminist attack on the male domination of society and the unequal distribution of power and resources to women may be justly motivated, but usually fails to point out that women, as well as men, have been responsible for this.

The female preference for wealthy, powerful, and aggressive husbands over poor, weak, submissive ones has been a potent force in the evolution of male character. David Buss writes:

Women today continue to want men who have resources, and they continue to reject men who lack resources. These preferences are expressed repeatedly and invariably in dozens of studies conducted on tens of thousands of individuals in scores of countries worldwide. They are expressed countless times in everyday life. In any given year, the men whom women marry earn more than men of the same age whom women do not marry. Women who earn more than their husbands seek divorce at double the rate of women whose husbands earn more than they do. Furthermore, men continue to form alliances and compete with other men to acquire the status and resources that make them desirable to women. The forces that originally caused the resource inequality between the sexes, namely women's preferences and men's competitive strategies, are the same forces that contribute to maintaining resource inequality today.

Likewise, male preference for youth and beauty in their wives has been the main driving force behind female competition and obsession over physical appearances. The powerful influence of commercial advertising in the fashion and cosmetic industries on reinforcing gender roles is not designed to suppress women, but merely reflects underlying male psychological preferences. It is not the fault of men that they are aggressive, domineering, and reckless, especially in competition with one another, just as it is not the fault of women that they spend so much time and money competing with one another to look more physically attractive. These are but the effects on one sex of evolved sexual preferences in the other. The social problems such gender related traits cause (and there are plenty) cannot be solved by blaming men or women for the way they are.

Balls

We have seen how sexual selection interacts with natural selection to power evolutionary change. We learned that the sex with less investment in child production and rearing is less choosy and more exhibitionistic, while the sex with more at stake tends to be very choosy. Sexual choice is constrained by divergent fitness distributions for the two sexes, with males occupying a broader spread.

Fitness distributions imply the existence of differences within and between populations. These differences are ultimately controlled by underlying gene activity. All individuals (except identical twins) are born with a different ensemble of genes, which account for most individual differences in things like body shape, hair color, and susceptibility to arthritis and heart disease. Genes also determine racial variations such as skin color and facial features. When it comes to human sexual differences, however, there is a very special gene that stands out.

All human embryos start out as tiny females. Their immature gonads, later to become sperm-producing testicles or egg-producing ovaries, are bipotential. But on a small stretch of the male Y chromosome in every cell of the male embryo, there is a very special gene called SRY. The presence of SRY in males is necessary to set in motion a cascade of genetic and biochemical events leading to maleness: the formation of testes, the male sex hormones testosterone and dihydroxytestosterone, sperm, and the constellation of physical and behavioral characteristics we associate with being male. Every cell of the normal female embryo has two X chromosomes and no Y chromosome. Therefore, females have no SRY gene. This results in the default formation of ovaries, female sex hormones, and feminine sexual characteristics.

The sex hormones testosterone, estrogen, and their relatives are produced by the fetal gonads and released into the bloodstream starting at about the eighth week of pregnancy. This has some effect on the subsequent growth and differentiation of the male and female fetus, but not much. It is hard to tell the sex of a newborn infant without examining its external genitalia. The major physical differentiation of the sexes coincides with a second burst of sex hormones some 12 years later, at the onset of puberty. But it is the first burst during fetal life that seems to 'prime' the brain in a sex specific manner. Testosterone and estrogen diffuse throughout the embryo's body and bathe the developing brain tissue, subtly but profoundly affecting the synaptic connectivity of billions of neurons in ways that science has only begun to discover. Infants may not be born 'little men and women' physically, but their brains have already been sexed.

Brains

There is evidence of anatomical differences between male and female brains. For example, women tend to have a larger corpus callosum. Males, on average, have larger hippocampi and a thicker cortex on the right side. They also tend to have larger brains per body weight than females. However, it is always difficult to determine the functional and behavioral significance (if any) of anatomical variations. What is clear is that men and women are exposed to different levels of sex hormones from before birth, which to some degree affects the way their minds work.

Neuropsychological experiments and surveys have repeatedly found women and girls to be better, on average, than otherwise matched men and boys on tests of language generation and comprehension (such as the SAT verbal section), fine motor coordination (knitting), arithmetic (bookkeeping), and visual item matching. In addition, girls were found to engage in more creative pretend play and show more empathy and cooperation than boys of similar age and education. Finally, there is much evidence that women have less 'lateralized' brain function then men, perhaps due to their greater connections between the two hemispheres, which could explain their better ability to 'multitask' and recover language function following left sided strokes. Men and boys are better, on average, in tasks of spatial manipulation (such as map reading and mentally rotating three dimensional objects), target directed motor skills (throwing and catching projectiles), and abstract reasoning (chess and higher mathematics). Male brains are also more lateralized than female brains, especially for language. The designers of many of these tests tried hard to correct for possible environmental factors such as different levels of exposure to certain activities and parental/peer encouragement of 'sex-appropriate' behaviors, but they still found statistically significant differences. Several studies on newborn infants have found that baby girls attend longer to social stimuli, such as faces and voices, while baby boys attend longer to nonsocial, spatial stimuli such as mobiles.

That sex-specific 'cognitive phenotypes' exist is no surprise. We've all heard the jokes about the male brain's hypertrophied modules for pornography, beer, dangerous sports, and channel surfing and the corresponding female brain's overactive modules for shopping, fashion sense, and commitment-seeking. The politically correct interpretation is that such differences are culturally acquired in a male-biased world. They are learned behaviors which may be difficult to change, but could just as easily have turned out the other way around in a different kind of society. However, this begs the question of why society turned out to favor these particular sex roles in the first place.

Most of us, even those who are critical of biological and cultural 'determinism', still dutifully live by most of these sexual 'stereotypes', consciously or unconsciously, and socialize our children to do the same. Cultural anthropologists have yet to find a culture in which females are generally better at throwing balls or reading maps, or males are generally better at compromising and cooperating. Historians have yet to describe a time or a place where it was the women who fought the wars, made the laws, and played the blood sports while men tended the house and children, prepared the food and spread most of the gossip. The simplest explanation for constant sex roles across different societies is that they have an underlying biological and evolutionary basis. This observation in no way justifies or condones sexual discrimination at the individual level. All individuals are unique and there is considerable overlap in abilities and inclinations between the sexes. Just as the tallest woman in the neighborhood is likely to be taller than most of the men around her, there are likely to be women who are better chess players than most men and men who are better social workers than most women.

I propose that biologically based cognitive differences do exist between men and women as a result of differential development of certain neural circuits, and that these differences have, to a large extent, led to the emergence of specific gender roles in society. The question is, how did such differences evolve in the first place?

Folk Physics and Folk Psychology

As we saw in earlier chapters, folk physics is the intuitive understanding of physical laws like causality, gravity, and geometry. Babies intuitively 'know' that objects are not supposed to float, move through solid walls, change direction without encountering another apparent force, and so on. Experiments with modern special effects have shown that they express 'surprise' (increased staring times) when shown such 'impossible events'. We all seem to be born with a built-in folk physics module.

Folk psychology is the 'common sense' acknowledgment of other free agents (other minds). It is the understanding of others' needs, desires, and emotional states independent of their physical state. Folk psychology is also innate. Children of two and three readily and spontaneously ascribe 'personalities' to cartoon dots and triangles moving around on a television screen, and four-year olds routinely use deception and pretense to obtain goals.

All of us are born with some potential degree of folk physics and psychology. Some of us (usually males) tend to be better folk physicists. Others (usually females) tend towards folk psychology. These differences start in childhood and are undoubtedly exaggerated by social and cultural factors. They become full-fledged after puberty as a result of further hormonal and environmental influence. Because not all males favor intuitive physics and not all females favor intuitive psychology, Simon Baron-Cohen uses the distinction 'male' and 'female-brain types'. Those fortunate individuals (of either sex) who are equally comfortable and capable of reading minds and understanding nature could be said to have a 'cognitively-balanced brain'.

The linkage between females and folk psychology on one hand, and between males and folk physics on the other, like the differential fitness spread we encountered earlier, may have its origins in a combination of natural and sexual selection. In terms of natural selection, males, who were already larger, faster, and stronger than females at the dawn of modern human history (the Pleistocene era), would benefit by developing a mind specialized for tracking moving game through the largely featureless African savannah. It would be useful to be able to judge distance, speed, and acceleration vectors, and to predict projectile behavior. In short, males would benefit by becoming better folk physicists. In contrast, females, who already tended to congregate in groups gathering edible tubers and berries and taking care of the more important 'domestic duties', would benefit by developing a mind specialized for assessing the needs and desires of their children, and for communicating and cooperating with other females. It would also be advantageous for females to become more adept at judging and choosing potential sexual partners based on their current and future potential as sources of material resources, physical protection, and good genes. Women would benefit by becoming better folk psychologists.

This natural selection explanation makes a good deal of sense, and probably contributed to the development of male and female-type brains. However, this was probably not the full story. A female with poor folk psychology is not any less fit to survive than a male with a similar degree of impairment reading minds; a male lacking intuitive physics is not necessarily more likely to die than a similar female. In Pleistocene society, perhaps even more so than today, people lived in tight social groups where cooperation and kinship support were as much the rule as competition and sexual division of labor. It is unlikely that the tribe would let a woman die simply because she had some trouble expressing herself or abandon a male relation to the lions just because he wasn't such a great hunter. More likely these traits were influenced by mate selection.

Sexual selection, like natural selection, cannot predict future utility. Evolution is, as the biologist Richard Dawkins put it, a 'blind watchmaker'. The purpose and design only seem apparent on hindsight. In humans, both males and females participate in mate selection. When men and women shop for potential mates, both generally try to find partners who are kind, trustworthy, intelligent, entertaining, and physically fit. True, men tend to emphasize youthful beauty, while women tend to concentrate on wealth and social status, but generally there is much overlap in the kinds of qualities both sexes look for. We do not consciously try to pick women for 'superior folk psychology' or men for 'superior folk physics'. But perhaps men with better hunting and navigation skills were attractive to women, or perhaps better hunters were able to dominate other males and win access to more (and more fit) females. Likewise, women with better communication and social skills may have been attractive to men, or perhaps more savvy women were able to out-compete other women for access to the more dominant men. In this way, sexual selection could have amplified the difference between the sexes already induced by natural selection.

Autism: an Extreme Male Brain?

It is interesting and often noted that the characteristics of the male-type brain are also traits commonly found, often in exaggerated and extreme form, in autism and AS. Autistic people, and AS to a lesser degree, have impaired linguistic, social, and multitasking ability. All of these are characteristics of the male-brain type. Additionally, autistics are often above average in mathematical reasoning, spatial manipulation tasks, and in finding small parts embedded in wholes. It is therefore attractive to think of autism and AS as pathological manifestations of an 'extreme male-type brain'. It is important to understand that possessing an extreme male brain is not the same as 'being extremely male' or having an overly masculine physique. Many people with AS are quite 'nerdy' and not very 'macho' at all. Autistic spectrum disorders are found in females (though in much smaller numbers) and thus have little to do with the presence of the Y chromosome. There is also no evidence that autistic males have higher levels of testosterone or other androgens. In fact, it has been observed that autistic and AS people are often effeminate, or otherwise psychosexually ambiguous. However, autism is up to nine times more common in males. There is clearly a link between male sex, the male-type brain, and the autistic phenotype, but it is a link that lies beyond chromosomes, hormone levels, or primary or secondary sex characteristics. Rather, it involves some complex interaction of defective genes which are normally involved in the sexual differentiation of the brain.

An aside on Race and Sex

I would like to clarify here that I am not gay. Despite this, I have occasionally been approached by members of the same sex making sexual overtures. I'm not sure if this occurs more frequently to me than to others, but it does seem that there is something about me that some men find attractive. One possibility is race. I am an Asian-American male in a predominantly non-Asian society. The presumably gay or bisexual men who have approached me have been either African-American or white, but not Asian.

I think that there is a sex-race stereotype in Western culture that covertly posits a sexual spectrum on different racial types. For both sexes, blacks are considered more 'masculine', whites are in the middle, and Asians (especially those from East and South East Asia) are more 'feminine'. Therefore, black men are considered the most masculine of males, and Asian women the most feminine of females, and by extension, black women are assumed to be somehow 'less feminine' and Asian men 'less masculine' than their white counterparts. This is a demeaning stereotype, but like most such stereotypes, it does have at least some small basis in fact. Sub-Saharan men, from whom most African-American men are descended, do have mildly higher testosterone levels than white men, who in turn have higher levels than Asian men. This explains, among a few other physical features, the higher level of prostate cancer (which is testosterone-dependent) in black men and the relatively sparse amount of body hair in Asian men. It may even explain the superior performance of black athletes in some sports. However, this is also probably responsible for the demeaning and unfortunately widespread stereotype of the 'over-sexed black man' and the 'submissive Asian woman'.

It is conceivable that the unwanted homosexual advances I've experienced are simply a reflection of prevailing racial stereotypes. Perhaps heterosexual Asian–American men experience disproportionately more of such advances than black men or white men in American society. I don't know. It is fairly clear that the rate of homosexuality has no racial variance. But it is also possible that males on the autistic spectrum, regardless of race, are more likely to be mistakenly thought to be gay and thus are more likely targets of homosexual propositions. If this is true, and I'm not aware of any studies on the matter, it may be related to the fact that autistic/AS people have problems with social judgment and are more likely to be socially naïve and sexually vulnerable. Most autistic people tend to be quite celibate, however, for that reason.

Why Don't Girls Like Me?

Most men with AS fail to find stable sexual relationships, and virtually all men with straight autism fail to have sexual relations at all. This is also probably true of women with AS/autism as well. If autistic and AS males have extreme-male type brains, why don't women find them especially attractive? Why did I have to be a 28 year old virgin? It was certainly not by choice. The reason for autistic men is quite clear. They are usually mentally retarded, and have very little communicative ability, which precludes the necessary social interactions for starting sexual relations. For AS, the situation is more complicated. While AS males also have communication and social difficulties, they often learn to compensate for them, and many are highly intelligent and productive members of society. The answer, I think, is that 'extreme maleness', whether in personality, physique, or the male-brain type, is not the most attractive for females. Women (and men) are attracted to proper balance and proportion. Just as men prefer a young woman with a 0.7 waist-to-hip ratio and properly symmetrical facial features over a freak with a 0.3 ratio or a fatso with a 1.2 ratio, women prefer men with a 'proper' ratio of intuitive psychology to intuitive physics. Men prefer women who are somewhat better folk psychologists, and women prefer men who are somewhat better mechanics. The extreme-male brain or extreme-female brain (if such a thing exists) is not preferred over a more reasonably balanced brain.

What's the Use of Autistic Genes?

Assuming that the autistic spectrum disorders are the result of specific combinations of defective genes and that those individuals tend not to have children, why haven't those 'bad' genes been selected out? The likely answer lies in the advantage that such genes may confer in 'lower doses' (ie. in those individuals who are heterozygous for one or more of those genes). There are many cases in which having one copy of a bad gene gives the individual immunity from some infectious disease or leads to some beneficial behavior which enables them to have more offspring.

For example, sickle cell anemia is caused by having two defective copies of a gene for hemoglobin, the oxygen-carrying protein in blood. People with this disease have blood cells that turn into elongated 'sickle' shapes under stressful or low oxygen conditions, producing severe pain, shortness of breath, and sometimes death from poor circulation and anemia. But individuals with just one defective copy (the sickle cell trait) are more resistant to infection by the malaria parasite, which normally attacks the healthy red blood cell and not the sickle one, causing weakness, extreme fever, anemia, and sometimes death. Thus the sickle cell trait has evolved and is quite prevalent in those parts of the world that have historically had high rates of malaria.

Similarly, the gene for cystic fibrosis, which produces the disease in people with two defective copies, appears to have conferred some degree of resistance in the heterozygous form to the tuberculosis bacterium and is thus commonly found in people whose ancestors used to live in areas where tuberculosis was prevalent. Schizophrenia and bipolar disorder are often linked to families in which artistic and intellectual creativity has been common. Often seeming defects are not defects at all, but rather evolutionarily successful adaptations gone awry.

Autism and AS are harmful genetic disorders of thought and behavior, but they are often found in families of physicists, engineers, and mathematicians. The existence of these so-called 'broader autistic phenotypes' suggests that there is some selective advantage in having 'autistic genes'. It is likely that these genes code for proteins that build brains to generate minds that think in certain logical ways to solve physical problems and attract women who admire those beautiful minds, producing children who are more likely to receive too many such genes and display the autistic phenotype.

◆   ◆   ◆

I am drifting in and out of a fitful sleep on the back seat of a tramcar. I shiver as the cold draft keeps waking me up. It is an early morning in Prague, April 1991. Mike Reznick and I are backpacking across Eastern Europe between terms at Oxford. With second-class Eurorail Pass and international youth hostel card in hand, we hit Berlin, Budapest, Dresden, Weimar, Wurtzberg, Vienna, Salzburg, and, of course, Prague, all Baroque gems of Prussian and Hapsburg high culture. Back in England, we all read Kundara's **The Unbearable Lightness of Being**. Now here we are in Czechoslovakia, just after the Velvet Revolution: the land of Kafka, fine Pilsner, and impossibly beautiful women.

The night we arrive, we check into a cheap hotel by the central station. Later we run into some hung-over Australian backpackers who advise us to check out a gaudy nightclub off Wenceslas Square. Mike, a tall thin art student and aspiring DJ from Manhattan, promptly gets busy making his rounds with a group of Danish school girls on their class trip. Meanwhile, I'm distracted by two platinum blondes grinding to a techno-beat dance track. I fantasize about having sex with these exotic women. Then one of them looks at me. The other one whispers something in her ear. I start to sweat.

'Yaah, Boooyyy!!'

I snap out of it.

'Which one?' Mike asks, with a wink.

'Huh?'

'Which one you want to fuck?'

'Uh, I'm not sure.'

'The tall one with the tight ass, for sure. She's staring right at you buddy! She digs you, man. I recommend you talk to her.'

'Okay', I reply in a daze.

I am exhausted and detached from reality. We aimlessly circle the streets of this hauntingly mysterious city for what seems like hours, until the first rays of the morning sun bathe us with surreal luminosity.

Part II: Genes

It is a truism to say that our individual identity is shaped by a complex interplay of genes and experience. We are, each of us, an amalgam of physical and psychological traits inherited from our two parents, and of things we have somehow picked up, knowingly or otherwise, during our lifetimes. But there is no clear dividing line between the end of nature and the start of nurture. The particular bodies our genes build affect how we subsequently interact with and learn from the world. A physically healthy, vigorous constitution may encourage outgoing behavior and inculcate leadership potential, while a shy introverted personality may lead to a life of scholarly achievement. Conversely, the environment in which we live and learn affects how our genes are expressed, and how our bodies and brains develop. We are biological organisms first, governed by a common genetic code, but all our genes and the proteins they make work within and are influenced by the world around them. We are really 100% nature and nurture. The genes set the general parameters for things like metabolic pathways, anatomical organization, and brain development. The actual outcome of these processes is determined by complex interactions with individually unique experiences.

Some diseases, like Huntington's disease or cystic fibrosis, are almost purely genetic. If you inherit the bad genes (one bad copy for autosomal dominant disorders like Huntington's, or two bad copies for autosomal recessive disorders like cystic fibrosis) you will get the disease, although the exact expression of the symptoms is somewhat dependent on other genetic and environmental factors. Other diseases, like appendicitis or AIDS, are almost totally non-genetic, although their course and outcome depend to some degree on immune status and cardiovascular health, which are themselves largely genetically determined. Most diseases fall somewhere in between on this gene-environment spectrum. Complex disorders such as autism, schizophrenia, obesity, or diabetes are caused by multiple defective genes and their protein products interacting with each other, with healthy genes and proteins, and with their environment.

In this chapter, we will concentrate on the genetic aspects of autism and Asperger syndrome, as there is ample evidence that they are heritable. The particular genes involved have yet to be found, but much exciting progress has been made recently, narrowing the search for possible candidates to several chromosomes.

Links

We are now ready to tackle the genetics of autistic spectrum disorders. Autism has long been known to run in families, though not at the high, predictable rates found for diseases caused by single gene mutations, like cystic fibrosis and Huntington's disease. The **relative risk** (**RR**) is a statistical value that measures the risk of disease occurrence in a family member relative to the risk in the general population. The RR gives a good indication of the extent of the genetic component in a disease. A RR greater than 2 indicates at least some degree of genetic susceptibility. The higher the RR, the greater the genetic component. Huntington's disease, which is autosomal dominant, has a RR of 5000. Cystic fibrosis, which is autosomal recessive, has a RR of 500. Schizophrenia and type 2 diabetes have RRs of 8 and 4 respectively. Autism's relative risk is about 100. The latter three diseases are 'complex genetic disorders', meaning that there are multiple genes as well as environmental factors sharing responsibility for the disease phenotype. Although each individual gene functions according to classic mendelian rules of segregation, the overall effect from the interacting factors complicates the inheritance pattern, hence the relatively low relative risk.

A better way of gauging the effects of genes is to study twins. Ideally, it would be useful to use dizygotic (fraternal) and monozygotic (identical) twins raised apart to eliminate the confounding environmental bias from growing up in similar environments. These are called 'adoption studies'. A disorder that is 100% hereditary (either autosomal dominant, autosomal recessive, or X-linked dominant or recessive) would be expected to result in a 100% concordance rate among the affected monozygotic twins. Autosomal dominant traits yield approximately a 50% concordance in dizygotic twins, while autosomal recessive traits yield 25% concordance. One twin adoption study of autistic children has revealed a monozygotic concordance rate of approximately 60%. [**Bailey, A., et al (1995) 'Autism as a strongly genetic disorder: evidence from a British twin study', Psychol. Med. 25, 63-77**]. This indicates that autism is not totally genetic. If it were, identical twins, who share all of their genes would have 100% concordance.

How, then, do we find the genes responsible for autism? There are several approaches one can take. The first is simply to visually inspect the chromosomes of autistic people in hopes of finding obvious defects such as breakages, inversions, or deletions of whole chromosomal segments. About 5% of autistic individuals do indeed have gross chromosomal defects. This does not mean that the defects are responsible for or even directly related to the autistic traits. Many of these individuals have a whole host of other problems in addition to autism, including mental retardation and physical deformities. However, one type of chromosome abnormality possibly related to autism does deserve attention. It involves a deletion of a region of chromosome 15 and is linked to two diseases called Prader-Willi and Angelman syndromes; we will return to this later.

A second approach is to study medical disorders with known genetic causes, which are linked to autism. Approximately ten to fifteen percent of autistics also have another genetic disorder. These include fragile X syndrome, tuberous sclerosis (TSC), and phenylketonuria (PKU). As many as 60% of those with fragile X and TSC are reported to have autistic traits. [**Folstein, S., and Rosen-Sheidley, B., 'Genetics of autism: complex aetiology for a heterogeneous disorder' in Nature Reviews Genetics, Dec 2001, 943-955**] But both disorders are quite rare and most autistics do not have fragile X syndrome, TCS, or PKU.

Unfortunately, the majority of autistic cases are idiopathic, meaning they are not related to any other particular defect or known cause. This forces us to use a third approach. We need to physically localize the actual susceptibility genes responsible for autism. Assuming that autism is linked to multiple defective genes spread randomly over 23 pairs of chromosomes, 30,000+ genes, and three billion bases of DNA, how do we even guess which barn has the proverbial needle in the haystack? In absence of better starting data, the best approach is to begin with a 'linkage screen' on a family pedigree with two or more autistic individuals.

In order to screen a family for autistic 'loci', it is imperative that we know what we're looking for. In other words, the autistic phenotype must be properly identified. Autism is a very heterogeneous disorder; it presents differently in different individuals. Intelligence, socialization, communication, and personality vary along wide and separate spectra from the most obviously autistic to more subtly disabled Asperger syndrome to odd but well-compensated individuals with the broader phenotype. Part of this may be related to genetic background effects (a single gene causing many different phenotypes in different individuals), but more likely, there is a great deal of polygenic influence (several genes contributing to multiple related phenotypes). There are types and subtypes of autism that may correspond to different underlying genetic backgrounds. For this reason, it is important that the subtypes of autism are identified, categorized, and segregated in family linkage and association studies to reduce 'background noise' from interfering characteristics. For instance, those with Asperger syndrome should probably be analyzed separately from those with more straightforward autism because they may well represent different populations, or even different (though closely related) disorders.

Once the phenotype definition is settled, we can then proceed to the next step: obtaining DNA from the affected individual and his/her relations. This is the easy part, usually involving a blood specimen for white blood cells (the red cells have no DNA) or a cheek swab for epithelial cells. The DNA from each individual is then analyzed to see which chromosomal regions the autistic individuals have in common.

Maps

All organisms need to reproduce their cells. In addition, all sexual organisms need to reproduce their gametes (sex cells). Asexual reproduction involves cell division in a process called **mitosis**. Here, the chromosomes are doubled and the cell simply divides. The daughter cells are genetically identical clones of the parent. Mitosis is how all asexual creatures reproduce. It is also how all somatic (nonsexual) cells of all sexual creatures such as fungi, flowering plants, fruit flies, and foxes reproduce. Basically, mitosis is how we all grow and regenerate tissue.

A distinctive feature of sexually reproducing organisms is a quite different and much more complex phenomenon called **meiosis**. Meiosis is the specialized process of chromosome reorganization and cell division of the early gametes, which give rise to the mature eggs and sperm. Unlike mitosis, meiosis involves the pairing of homologous chromosomes (recall that humans have two copies of each chromosome, one from each parent) and usually at least one process of chromosomal recombination per chromosome. Recombination is a physical breakage of the DNA from each homologous chromosome, which then becomes joined to the corresponding part on the sister chromosome. Part of the resulting recombinant chromosome will have the genes the cell inherited from the mother, while the other part will have genes from the father. The chromosome itself is a novel blend of mom and dad.

The eggs and sperm are thus full of naturally recombinant DNA. All the genetic information in the resulting daughter cells of all individuals of all species of sexual life on earth are not just combinations of maternal and paternal chromosomes, but also products of chromosomal recombination events that happened in all their ancestors going back a billion years. Recombination provides the raw variety and random variation that allows some offspring to survive the vicissitudes of harmful parasites, mutations, and other adverse factors of nature.

◆   ◆   ◆

The key to mapping genes was the discovery of genetic polymorphisms. Polymorphisms are sites of DNA that differ between individuals. They can be functionally important, corresponding to different alleles of a specific gene, or they may be insignificant, simply different spellings in the mass of junk DNA. But either way, polymorphisms can be localized to relatively precise locations in the genome, such as a small part of the short arm of chromosome 7 or the long arm of chromosome 4, for instance. And because different people have different polymorphisms, they make good markers.

If we discount the presence of chromosomal recombination during meiosis, a disease gene will **segregate independently** of a marker located on a different chromosome, but the two will **segregate together** if they are on the same chromosome. Using polymorphic markers on each of the 23 pairs of chromosomes, we can analyze the DNA of members of a family with two or more autistic individuals to see which markers segregate with the autistic trait. This localizes the gene or genes to a particular chromosome.

Next, we use the concept of 'linkage analysis' to zoom in on the particular region of the chromosome where the gene is likely to be found. Linkage analysis depends on the knowledge of meiotic recombination: the closer a trait is physically located to a marker on the chromosome, the more likely they are to segregate together. However, the further apart they are, the more likely it is that they would become separated by a recombination event. Analysis of recombination frequencies can give a fairly good approximation of the physical location of target genes. In fact, late one night in 1911, a 19-year-old Columbia University undergraduate named Alfred Sturtevant got the idea to use the frequency of recombination between different visible traits in fruit flies to 'map' the genes responsible for these traits relative to one another. Sturtevant had constructed the first genetic map. He christened the unit of genetic distance 'centimorgans' in honor of his teacher and a founder of modern genetics, Thomas Hunt Morgan. The distance in centimorgans between two genes is the percentage of times that they are separated by recombination. Anything less than 50 centimorgans indicates that the traits, genes, or markers are linked to the same chromosome. The closer the linkage, the less frequent the recombination, and the shorter the distance in centimorgans. One centimorgan in human DNA corresponds roughly to one million base pairs (1 MB).

There are several types of genetic markers. One commonly used in the past, though now going out of style, is the 'Restriction Fragment Length Polymorphism' or RFLP. RFLPs refer to DNA sequences which are slightly variable in different people resulting in different patterns of products when cut at specific points or 'restriction sites' by special bacterial enzymes called restriction endonucleases. For example, one well-known restriction endonuclease, called 'EcoR1', derived from the E. coli bacteria, recognizes and cuts DNA only at the sequence GAATTC. Another enzyme, 'HindIII', from the Hemophilus influenza bacteria, cuts only at the sequence AAGCTT. The discovery in the 1970s of restriction endonucleases and another type of bacterial enzyme, DNA ligases (which reattach loose pieces of DNA), has made the revolution of recombinant DNA science possible.

Another useful type of genetic marker is the so-called 'microsatellite tandem repeat', which is a short stretch of DNA (two to four base pairs long) that is repeated dozens or hundreds of times in a given individual, but with each individual having a unique number of repeats. Microsatellites are located at hundreds of thousands of locations throughout the genome. Finally there are 'single nucleotide polymorphisms' or (SNPs), which are simply random sites in (usually noncoding) DNA which have a different base substitution in different people. There are millions of SNPs, occurring on average about once every kilobase (1000 base pairs) or so, and they thus allow very good localization.

Once the gene has been localized to within several hundred kilobases using ever-finer linkage marker analysis, one can examine the human genome sequence to see what genes are predicted to be located in that interval. One can then search for specific mutations in any interesting candidate genes.

Scientists have used linkage analysis of autistic families to narrow down the location of possible autism causing genes. The strongest and most consistent signal collected from several recent genome screens comes from the long arm of chromosome 7 (7q).

7

Once the linkage studies localized autism to markers on chromosome 7, specifically the 7q22-31 region, scientists started to look for candidate genes in the area. 7q22-31 spans 10 centimorgans or about 10 million base pairs, still a relatively large region. In this stretch of DNA, there are several important genes including, incidentally, the 'Cystic Fibrosis Transmembrane Receptor' (CFTR) gene, which, when mutated, causes cystic fibrosis. Another gene is called RELN, which encodes a protein that is important for neuron migration. Individuals with a mutated RELN gene develop a disease called autosomal recessive lissencephaly, where the layers of cells in the cerebellum fail to form properly, resulting in neurological problems.

Another gene on 7q is called WNT2. It belongs to a family of genes that encode the so-called 'wingless' proteins, which are important in regulating embryonic development. The wingless protein is part of a developmental regulatory pathway that was originally discovered, along with so much else in molecular genetics, by fruit fly scientists. Fruit flies are ideal (model) organisms because they share many of the same genes and proteins that 'higher organisms' like mice and monkeys use in embryonic development, but are quite cheap and easy to manipulate in the lab. In fruit flies, the wingless protein is part of the 'wingless signal transduction pathway'. Wingless is a secreted protein that is produced by certain cells of the developing embryo, diffuses some distance away, and activates its receptor on the surface of other (target) cells. When wingless binds to its receptor protein, called 'frizzled', anchored to the target cell membrane, frizzled changes its shape, and activates a series of other proteins further 'downstream' in the pathway. The last such protein in this 'cascade' enters the nucleus of the target cell and activates a transcription factor. Transcription factors bind to certain sites on select genes and activate them. The specific genes turned on depend on the properties of the target cell, rather than on the incoming wingless signal. The wingless protein is responsible for the organization or 'patterning' of many tissues in the developing embryo, including the body segments, the eye, and the wings of the fruit fly, and much of the vertebrate nervous system, by differential regulation of tissue-specific genes. Some embryonic cells will become part of the arm, some will become part of the eye, and some will end up parts of the brain, thanks to the action of wingless and other signaling pathways. Scientists have isolated and cloned many genes involved in developmental signaling, including WNT, RELN, and 'disheveled' (which encodes a downstream protein in the wingless pathway). They have recently been able to engineer defective or 'knockout' copies of these genes and then insert them into mouse embryos to produce adult mice with specifically deleted genes. Knockout mice have been made for the RELN, WNT, and disheveled genes. WNT knockouts are embryonic lethal. RELN knockout mice were found to have mobility and coordination problems much like in human lissencephaly, and similar, perhaps, to some aspects of autism. The disheveled knockout mice were characterized by reduced social interaction (lack of normal huddling during sleep, mothering behavior, and grooming) again similar to some aspects of human autism.

Perhaps the strongest candidate thus far for an autistic locus on 7q follows the discovery of the FOXP2 gene. The story starts with a curious disorder of speech and language processing first described in 1990 in an English family, the KEs. The disorder, now called 'Specific Language Impairment' or (SLI) affects both the comprehension and generation of spoken and written grammar. In the KEs, the trait is inherited in an autosomal dominant pattern. We will return to the linguistic aspects of SLI later. In 1997, a team of geneticists from Oxford University used linkage analysis to localize the putative SLI gene to a small region of chromosome 7q. The gene itself was subsequently isolated, cloned, and sequenced. The gene is called 'Forkhead box P2' or FOXP2 and is a member of a family of transcription factors. Transcription factors are quite common in the human and other genomes, comprising perhaps 10% of all genes. In the KE family, the FOXP2 gene was found to have a mutation of just one nucleotide (called a 'point mutation'), but it is enough to render the protein defective. Children born with just one copy of the mutation suffer abnormal brain development, especially involving the speech and language centers on the left parietal lobe, which can be picked up on MRI scans.

SLI and autism are separate disorders, but the co-occurrence of the two in affected families is considerable. Autism has been found to have a prevalence of 3% in siblings of SLI individuals, compared to 0.1% in the general population. In addition, language disorders of various types are quite prevalent in families of autistic individuals. As we shall see in chapter 8, autism itself is in many ways a disorder of communication. About 25% of first-degree relatives of autistic children have delayed onset of speech or reading problems as compared to 5% to 10% in the general population.

FOXP2, the gene responsible for SLI, does not by itself cause autism. But given the strong genetic linkage between autism and this region of the genome, and between SLI and the autistic phenotype, it is likely to play an important role. It is very possible that FOXP2 may be one of the 'autistic genes'.

15

Earlier, we explored the possibility that autism may be a pathological exaggeration of 'normal' male/female psychological differences; that perhaps a combination of sexual and natural selection has created greater male vulnerability not just for the autistic phenotype, but also for many other developmental disorders, and also for the broader phenotype that we often see in mathematicians, engineers, and scientists. Now we will take a closer look at the possible genetic mechanism behind these sexual differences, and, at the same time, look for more 'autistic genes'.

Our next stop is chromosome 15. It has not been as strongly implicated in linkage studies as 7q, but it is nonetheless important for two other reasons. First, chromosome 15 is the most common site for chromosomal rearrangements in autism (although most cases of autism are not associated with apparent chromosome rearrangements). Second, it is a major source of 'genomic imprinting' in humans.

Ordinarily, all human genes come in two copies, one from the maternal and the other from the paternal coterie of chromosomes. The two copies, or alleles, may be the same (homozygosity) or different (heterozygosity). Both copies of most genes are usually used to make protein. For some genes, however, only one copy is used. This is called 'allelic exclusion'. These include all the genes on the X chromosome in females (we will return to this concept later), the immunoglobulin and T-cell receptor genes of the immune system, and the olfactory receptor genes, which code for the various odor receptor proteins. Recently, another mechanism of allelic exclusion was discovered: genomic imprinting. It is a fascinating process by which one copy of certain genes is always shut down depending on the sex of its parent of origin. For some genes, such as insulin-like growth factor type 2 (IGF2), only the paternally acquired copy is expressed. For other genes, such as the ubiqutin ligase E3A (UBE3A), it is the maternal copy that is turned on.

Very basically, genomic imprinting works as follows. Genes exist on DNA which is normally tightly wound around proteins called histones. This DNA/protein complex makes up the chromosomes. For the DNA to be transcribed, it must first be unwound and 'opened-up' for the various bulky transcription factors to get to work on it. This involves the addition of certain chemicals, called acetyl groups, to the histones, and the removal of other chemicals, called methyl groups, from certain cytosine residues on the DNA. When DNA is methylated, especially on its promoter regions, transcription cannot occur and the gene is effectively shut down. Likewise, when the histones are free of acetyl groups (deacetylated), the chromosomes are tightly wound and again no RNA is transcribed. In the germ line cells (eggs and sperm) prior to fertilization, there is little need for gene expression, so most DNA is methylated and histones, deacetylated. Immediately after fertilization, however, there is a tremendous amount of gene activity as the embryo develops and cells rapidly divide. This requires the activity of demethylation and acetylation enzymes throughout the genome. However, not all genes are demethylated (turned on) in the developing embryo. Some genes retain an 'imprint' from their gametes. For a maternally imprinted gene, only the maternal copy in the egg will be turned on. The paternal copy from the sperm will remain methylated. The reverse is true for a paternally imprinted gene. This occurs regardless of the sex of the offspring. But, importantly, **this sex-specific imprint will be erased and re-imprinted on the newly developing germ cells of the embryo**. For example, a maternally derived, maternally imprinted gene will be kept on in her son's body, but will be shut off in all of the son's sperm. A paternally derived, paternally imprinted gene will be kept on in his daughter's body, but will be shut off in all of the daughter's eggs. There is a new imprint put on these special genes for each pass through the germ line.

Imprinting occurs in some flowering plants, insects, and all placental mammals, but not in marsupials (like kangaroos) or in monotremes (egg-laying mammals like the duck-billed platypus). Most genes are not subject to imprinting, but many of those that are seem to affect fetal growth and development. For instance in mice, IGF2 (insulin-like growth factor 2), which is paternally imprinted, tends to increase fetal growth, while the IGF2 receptor, which is maternally imprinted, tends to counteract the effect of IGF2. An interesting hypothesis is that imprinting evolved as the result of 'sexual conflict' between the parental genomes (as opposed to conflict between male and female individuals): the paternal genome propagates itself best by creating an embryo which aggressively removes nutrients from the mother; the maternal genome suppresses this to protect the mother and spare some resources for future offspring.

In human chromosome 15q, there is a major imprinting center that is sometimes involved in chromosome breakage events in either the maternal or paternal chromosome. These result in one of two distinct and clinically fascinating disorders. One gene, called SNRPN, encodes a protein involved in mRNA splicing. It is paternally imprinted. If the chromosome breaks in this region of the **paternal chromosome**, leaving only a non-expressed maternal copy on the other chromosome, the child will suffer from Prader-Willi syndrome (PWS). Males and females with PWS are characterized by mild mental retardation, poor language, loss of motor tone, and a bizarre preoccupation with food. These children will eat constantly, usually becoming morbidly obese. Just adjacent to this PWS locus is another gene, the maternally imprinted UBE3A mentioned earlier. If a deletion occurs in this region of the **maternal chromosome**, the affected boy or girl will develop Angelman syndrome, characterized by mental and growth retardation, hyperactivity, and wild and inappropriate laughter. There may be some association between PWS/Angelman syndrome and autism, and additionally, a few linkage studies have implicated a nearby region possibly involving genes for the receptor for the neurotransmitter, GABA. We will return to the role of neurotransmitters in the pathogenesis of autism later.

X

We now turn our attention to the X chromosome. Given the male preponderance of autism, Asperger syndrome, the broader autistic phenotype, and so many other behavioral and developmental problems, it is a rather logical place to look for autism causing genes. Genome scans have not yet revealed much evidence of genetic linkages to X. But there is some remarkable evidence from another disorder, Turner's syndrome, for a possible imprinted locus somewhere on the X chromosome predisposing to autism.

The X and Y-chromosomes are unique in that they are not homologous. The X has plenty of genes, but the Y has very few aside from the very important male-determining SRY gene. Both sexes need the genes on the X, but only males need the Y. So the Y evolved to stay in males and avoid recombination with the X. The problem is that females have two Xs, causing a potential overdose of gene expression that our ancestors found disadvantageous. The solution was to shut down one of the two Xs at random. This is what happens.

A female receives an X from each parent. A male receives an X from his mother and a Y from his father, and the maternally acquired X is turned on as usual. In the female, the presence of two Xs seems to activate the expression of a unique gene, called 'Xist', from one of the X-chromosomes. The Xist mRNA coats the entire X-chromosome and causes it to condense via DNA methylation and histone deacetylation, thereby inactivating it.

The X to be inactivated is chosen randomly at an early stage of development. Thus those cells in which a maternal X is turned off will give rise to paternal clones in all of their progeny, while those cells with a paternal X inactivated will give rise to maternal progeny. The resulting female offspring will be a 'mosaic' or mixture of cells, some of which have a maternal X, and some of which will have a paternal X. Female calico cats have a gene for coat color on their X-chromosome. These female felines, being X mosaics, end up with the familiar mixed calico pattern.

It was recently discovered that some genes are actually imprinted on the X-chromosome. In other words, there are genes expressed only from a maternally or paternally acquired X regardless of which X is subsequently silenced in the female. The English developmental geneticist David Skuse has conducted an elegant series of experiments on Turner's syndrome individuals to find out if X imprinted genes are related to differences in neuropsychological profiles. Turner's syndrome is a female chromosomal disorder in which one of the X-chromosomes from either parent is lost very early in development. Thus girls with Turner's have only one X. In many of them, **all the cells in their bodies will have the same X**. In 70% of these 'monosomic' Turner's, the X is maternal in origin, and in 30%, the X is paternal.

Turner's females tend to have short stature and thick necks. They are usually infertile, have low-normal IQs, and often display learning disabilities and social incompetence. Dr Skuse separated a group of 79 Turner's females aged 6 to 25 with maternally derived Xs, from 31 females with paternally derived Xs. He then gave their parents a questionnaire to assess the subjects' social communication competency relative to age matched normal males and females and to autistic subjects. He found that normal females scored an average of 2, normal males 4, paternal X Turner's 5, maternal X Turner's 9, and autistic subjects 16 (out of 20).

The study makes it clear that there is something imprinted on the X-chromosome which adversely affects the social and communicative behavior of individuals with a maternally inherited X. In addition, there seems to be a gradient in social competence with females>males>paternal X Turner's>maternal X Turner's>autism. This is somewhat akin to what we said about folk physics vs folk psychology earlier. Skuse postulates that this locus is paternally imprinted because males (who inherit their single X from their mother) are more vulnerable to autism than females, who receive the presumably 'protective' X from their fathers. He adds, 'the actions of the locus may not be to increase liability to autism as such but rather to increase male vulnerability to social communication impairments in a range of neurodevelopmental disorders, which preferentially affect males.' According to his 'imprinted-X liability threshold model', the X-linked locus lowers the threshold for other autism causing genes on other chromosomes to contribute to expressing the autistic phenotype. Normal males, by inheriting only one X (the 'silent imprint'), are already at a disadvantage relative to normal females. Turner's females (both types) are at a higher risk because of generally decreased X expression, called 'haploinsufficiency'. Maternal X Turner's are worse than paternal X Turner's due to their silent imprinted locus. Those with autistic genes on various other chromosomes such as 7q are obviously at higher risk, and being a male with these bad genes compounds this risk even further.

Skuse's model goes a long way to explaining some of the conflicting and confusing data collected thus far. If autism is a polygenic, multi-chromosomal disorder affected by an imprinted X-linked locus, there would be no clear sex linkage inheritance pattern, which is what we find. The traits that are better developed in normal females, such as socialization and communication, as opposed to male-type traits, such as spatial manipulation and local coherence, should also be less deficient in autistic females as compared to autistic males. In many ways, they are. There is evidence that autistic females tend to be somewhat better socially and worse mechanically than their male counterparts with similar IQs. Low IQ, in general, may lower the threshold for the expression of the autistic phenotype **independently** of any protective effect of the paternally imprinted X locus. Thus we should expect to see a less extreme male/female ratio in retarded autistics as compared with more intelligent autistics such as those with Asperger syndrome. This is indeed the case. The male: female ratio is four to one for those with low IQs, but greater than ten to one for those with Asperger syndrome and high functioning autism.

The imprinted X locus explanation for autistic liability is intriguing; but even if it is true, it does not solve the entire genetic puzzle of autism. There are undoubtedly many autistic genes scattered over the three billion base pairs of DNA, and they and their protein products have complex interactions affecting the development of the brain. Molecular genetics will help us find these genes. But finding the genes is only the next step. For those with the vision to see, developmental neuroscience will take us further in our quest, from genes and proteins to a cacophony of buzzing neurotransmitters to an elegant universe of dancing axons and neurons inside all of our heads.

◆   ◆   ◆

Golden light filters into the Upper East Side Irish pub on a late September afternoon in 1998. It is warm and pleasant, just like the afternoon when I saw Nora off at the airport. But the very thought brings a weight on my chest. It is time for another drink. I guess you can say I've been a regular here for the last few months; alcohol and the comfort of being around strangers ease the loneliness and doubt, if only for a while.

There is a new bartender today. He doesn't know me, so there's no pressure to act normal. Plus, the two pints of lager and repeated rehearsals lubricate my behavior:

'So may I ask what county you're from, my good man?' I ask, instantly recognizing his Irish features before he even opens his mouth.

'Roscommon,' the bartender replies with a characteristic lilt.

'A few tics south of Donegal,' I nod.

'You know Ireland?' he asks with a gentle look of surprise.

'I used to know a girl from Killybegs, but she's gone off to work in London now,' I say dejectedly. 'Went to visit her last month, but it appears she's getting along with her own life.'

'Life goes on, and it's probably just as well. They got a temper about them, you know, these Irish girls, believe me,' he says.

We banter around for a while until the bar starts to get busy. I tell him of my infatuation with Irish and British accents and with young women from the British Isles. Fortunately, I don't reveal that I was hoping to lose my virginity to this woman before my upcoming 28th birthday.

The New York evening deepens, and a loud and youthful crowd starts to settle in around me. I seem to be surrounded by attractive young women talking and laughing with one another. They occasionally glance over at groups of young men intently watching sports on the overhead television. Occasionally, a lone male or a group of males walks over to some women, and starts making animated comments about something or other. It is impossible to hear what's being said, but I see the body language. The male and female components slowly and almost imperceptibly inch up towards each other while gesticulating.

Several beers later, I decide to take the same approach myself. I see two young ladies talking a few feet in front of me. The petite one with short blond hair is wearing a tight black skirt and heels. The other one, also very slim with a great figure, has darker features, perhaps Persian or Indian, with large round eyes and straight brown hair. She is wearing a blouse unbuttoned to reveal the top of her breasts. I walk up to the women and introduce myself (like an idiot): 'What's up tonight? Pretty good music, don't you think? This is a really cool bar and I know the bartender, he's pretty cool, too.'

The two stop talking, look at each other and smile. They are beautiful. The sexy blonde leans close to me until I can smell her perfume. She asks me where I'm from. Before I can reply, the Indian looking woman whispers something to her, then tells me they have to go to the ladies' room. I wish I knew what they are thinking. They turn away from me and continue laughing and sipping from their cocktail glasses. A few minutes later I see them talking to a couple of young investment banker types in suits. I decide I've had enough, but before I can finish my beer and leave, a plump middle-aged man puts his hand on my shoulder. I jump.

'Don't worry about it! These girls are tough. It happens to everybody!' he slurs through bad teeth, his breath reeking of whisky and cigarettes. 'You know, I've seen you around a couple of times. You live in the neighborhood?'

I hesitate, 'uh, uh yeah, on Seventy Fifth Street between Second and Third,' I tell him without thinking. I could kick myself.

'Well how about that? I live just block away from you,' he says with a diabolical smile. 'The name is John, but you can call me Jack. Can I get you another drink?'

Part III: Development

At first glance, there is nothing particularly unusual looking about the autistic brain. There are no gross abnormalities, tumors, or cavities. Even under the microscope, it is difficult to find any consistent pathology common among cases of autism. There are no telltale signs such as the characteristic absence of dopaminergic neurons of the substantia nigra one finds in Parkinson's disease or the amyloid plaques associated with Alzheimer's disease. But if we look closely, we can find abnormalities. Lots of them. Some autistic brains are unusually large [**Piven J, et al (1995) 'An MRI study of brain size in autism'** American Journal of Psychiatry 152, 1145-1149]. Others have asymmetrical lesions in the right parietal lobe, or the left temporal lobe, or vice versa [**McKelvey, et al (1995) 'Right hemisphere dysfunction in Asperger syndrome'** Journal of Child Neurology 10(4) 310-314] & [**Jones, P & Kerwin (1990) 'Left temporal lobe damage in Asperger syndrome'** British Jour of Psychiatry 156, 570-572]. Some have missing or underdeveloped portions of the cerebellum

[Courchesne, E. et al (1988) 'Hypoplasia of cerebellar vermal lobules V & VII in autism' <u>New England Journal of Medicine</u> 318, 1349-1354]. Others have abnormalities in parts of the brainstem [Rodier, P.'Early origins of autism' <u>Scientific American</u> Feb, 2000]. Still others have lesions of the corpus callosum [Berthier, ML (1994) 'Corticocallosal anomalies in Asperger syndrome' <u>American Journal of Roentgenology</u> 162, 236-237]. On histological (microscopic) examination, even more abnormalities become evident. Some brain tissue samples have revealed evidence of unusually small Purkinje cells in the cerebellum; others show decreased dendritic branching in the limbic system [Bauman, ML & Kemper (1985) 'Histoanatomical observation of the brain in early infantile autism' <u>Neurology</u> 35, 866-874]. If you're really out to find differences, you are likely to find something. The problem is lack of consistency and evidence of causality. How do we know which anatomic defect, if any, is actually causing the autistic phenotype? A particular finding, such as an absent cerebellar vermis or small facial nucleus **may be associated** with autism, but that doesn't mean that one caused or resulted from the other. Even if we can prove that a particular histological or anatomical abnormality results in autism in certain cases, what about all those other cases of autism that are not associated? Additionally, if the lesion is large enough, or involves an important structure, such as the brainstem or thalamus, the end result may be autism, but along with a dozen other neuropsychiatric defects. To say that a brainstem lesion 'causes' autism is like saying that irreconcilable differences 'cause' divorce. These explanations may be true, but are not very informative.

◆   ◆   ◆

So where do we start? We can set up a few ground rules. **First, autism is highly genetic.** Earlier, we discussed a few chromosomal areas that may have candidate genes highly linked to autistic families. These regions are excellent places to start looking for genes that increase susceptibility to autism. We have already mentioned a few and will introduce several more.

Second, autism is a developmental disorder. As such, it tends to follow a specific time course. Affected children fail to meet the age appropriate developmental milestones of interactive play, spontaneous pointing and speech, and goal-directed gaze. There may well be less apparent deficits occurring as early as 12 to 18 months of age. Certainly by five years of age, autistic behavior is clearly evident. So we should look for factors that affect early neurological development.

These may be genetic, aspects of the prenatal environment, or experiences in infancy.

Third, autism has certain consistent psychological characteristics. These allow it to be distinguished from other disorders such as schizophrenia, obsessive-compulsive disorder, or attention deficit disorder. As we have seen in earlier chapters, these characteristics involve abnormalities in higher-order perceptual integration, motor control, emotional regulation, imagination and memory, flexible decision-making, and the understanding of other minds. In addition, as we shall see in the next chapter, autistic people have serious problems with communication (verbal and nonverbal). We already have a fairly good understanding of **where** many of these protean processes occur in the normal brain. It is thus logical to examine these areas and neural circuits in autism.

These three rules give us some idea about **how** autism is transmitted, **when** it is expressed, and **where** it might occur. Any theory purporting to 'explain' autism should, at the very least, be consistent with the existing scientific knowledge about the genetics, developmental neurobiology, and functional neuroanatomy of the autistic disorders. An explanation, however clever, which accounts for the social and communicative deficits in terms of the underlying anatomical abnormality is incomplete if it does not address the process of neurodevelopment. Likewise, a developmental model, however successful in explaining the autistic phenotype, is likely to be flawed if it fails to address the genetic aspects of autism. In this chapter, I will attempt to construct a developmental model of autism. This model will build largely on the recent genetic work that has yielded many intriguing discoveries. In the end, I hope to explain how certain genetic factors, in combination with environmental triggers, produce developmental cascades leading to the constellation of deficits we know as autism.

Synthesis

Earlier, we found that autistic individuals have problems in high-level perceptual processing. They lack the 'central coherence' to integrate bits of sensory input and self-generated impulses into a global picture. This failure to get the gist and make common sense of everyday situations is the basis for much of the social and communication difficulties autistic individuals constantly face. The anatomical foundations for the defective central coherence may be a dominant left cerebral hemisphere, or an abnormality in interhemispherical communication via the corpus callosum. As we have noted, there is evidence of callosal thinning in some autistic subjects. Another possibility is defective contact between the thalamus and the cerebral cortex. The thalamus connects and integrates much of the 'higher-level' cortical activity with 'lower-level' subcortical activity. This connectivity starts to develop well before birth, but gradually undergoes selective 'pruning' as the child's brain adapts to its world. Billions of connections and neurons are eliminated in this normal developmental process. We know that many autistic children have relatively large brains and heads. It is possible, then, that there is a problem with pruning, causing an excess of neurons and non-specific neural connections, especially between the cortex and the subcortical areas. Perhaps this excess 'noise in the system' leads to a loss of central coherence. We will return to this important point later.

In chapter 3, we examined the motor abnormalities in autism, and speculated that they may be linked to a defective system for monitoring the voluntary movements of others. Motor programs can be divided into 'automatic routines', coded in the subcortical structures of the cerebellum and basal ganglia, and conscious actions, coded in various parts of the frontal and parietal cortices. A dissociation between these areas may result in corresponding disconnection between spontaneous (usually intact) and planned (usually clumsy) movements we often find in autistic individuals. Anatomical studies have not revealed major defects in the basal ganglia, but there are reports of cerebellar abnormalities at both the gross and microscopic levels.

Chapter 4 addressed the affective characteristics of autism. There is a curious mixture of social detachment and indifference, on one hand, with a more 'primitive' emotional lability and anxiety, on the other. Part of this relates to an inability to assess and understand nonverbal facial gestures. Functional MRI experiments on autistic subjects viewing pictures of faces demonstrate activity in the inferior temporal gyrus rather than in the fusiform gyrus, the typical 'face processing area'. Autistics seem to view people more like objects, and treat them as such.

Perhaps more importantly, the amygdala is dysfunctional. The amygdala is highly connected to the temporal and the frontal brain areas, and is responsible for supplying an 'emotional valence' to experiences, memories, and thoughts. It modulates the activity of the cortex, but is also normally held in check by cortical activity. In the case of autism, there seems to be a disconnection between this subcortical source of raw emotion and the higher cortical regions where the emotions are transformed into **feelings**. This is where the generalized anxiety and flash outbursts of temper common to autism are likely to arise. Histochemically, there is evidence that the neurons of the amygdala and other parts of the limbic system are smaller and abnormally packed. This localized pathology can nonetheless have very widespread developmental effects. We have seen in monkey models that early damage to the developing amygdala affects dopamine activity in other brain regions, especially the PFC.

Finally, in chapter 5, we explored the constellation of behavioral abnormalities ranging from simple perseveration and motor stereotypies to failure of executive function. What they all have in common are deficits in attentional control, likely mediated in thalamo-cortical circuits. Parts of the frontal cortex (the dorsolateral PFC and the anterior cingulate) integrate lower-level sensory and motor information to control planning and decision-making. In addition, other parts of the frontal cortex (the orbital and medial PFC) are linked to the amygdala and temporal areas and send social and emotional signals to the decision-making areas. Neuroimaging studies, such as those of Fletcher and Happe cited in chapter 4, have found derangements of normal frontal activity in high-functioning autistic subjects engaged in TOM story analysis. Other studies on autistic children have found evidence of decreased metabolic activity in the PFC [**Ernst, M et al (1997) 'Reduced medial prefrontal dopaminergic activity in autistic children' Lancet 350, 638**] and in the anterior cingulate [**Hazendon, M et al (1997) 'Anterior cingulate gyrus volume and glucose metabolism in autistic disorder' American Journal of Psychiatry 154(8) 1047-1050**]. These studies support the idea that autism is a disorder of the emotional and attentional aspects of decision-making. To quote the autism researchers Robert Schultz, Lizabeth Romanski, and Katherine Tsatsanis [**Schultz, et al (2000)'Neurofunctional models of autistic disorder and Asperger syndrome', in Klin & Volkmar, Asperger Syndrome pg194**]:

'The orbital and medial PFC may have a specific role in integrating affective with cognitive processes. Studies of the orbital cortex in patients and nonhuman primates suggest that it participates in integrating information about rewards and punishments to bias future behavior. The integration of social and emotional information is critically important vis-à-vis the primary deficits found in [autism and Asperger syndrome]. Because the orbital and medial PFC are densely interconnected with limbic areas, especially the amygdala, it may be more appropriate to conceptualize a limbic-frontal system as critical to [autistic] symptomatology, as opposed to isolated regions in one or the other. This pattern of connections would allow the amygdala to transmit information about emotional significance to prefrontal regions involved in guiding behaviors such as social interactions and related social cognitive processes. Disconnection of the amygdala from orbital and medial prefrontal cortical processing regions could result in the failure to transmit information about stimulus valence and also a more general dysregulation of emotional responsivity. In this scenario, social stimuli would not acquire their normal valence, causing social processing difficulties similar to the insensitivity to consequences displayed by patients with acute ventromedial prefrontal lesions.'

◆   ◆   ◆

What we need to do is find some common ground among the mountains of results from the behavioral, anatomical, cellular, and genetic studies. Furthermore, the model must be consistent with what we already know about neurological development. This is a daunting task. Assuming that autism is a neurochemical, genetic, and/or cellular level defect expressed in certain areas of the brain at certain periods of development, '...we would be studying neurochemicals at the level of single synapses with at least millisecond temporal resolution over the course of development from fetal life to at least age 5.' [Cook, E (1996) 'Pathophysiology of autism: neurochemistry' **Journal of Autism and Developmental Disorders** 26(2), 221-225].

But things are not as hopeless as they may seem. We can start to simplify this complex picture. First, we know that quite distant and distinct regions, such as the amygdala and PFC, the motor cortex and cerebellum, or the brainstem nuclei and the extraocular muscles of the eye, often enjoy intimate functional and anatomical relationships. Even at early stages of development, these distinct areas communicate with each other. Secondly, functional and behavioral pathologies that seem very different, such as repetitive motions, defective TOM, and poor executive function, may actually be different manifestations of a common developmental defect, such as attentional failure. Finally, developmental problems that manifest years apart, such as a failure to point and a lack of social interaction may be ramifications of a single latent abnormality. In fact, early defects in social or emotional behavior can themselves prevent the normal development of social cognition, even though this may not be apparent until much later.

As tempting as it is to simplify this picture, we must not err in oversimplifying it. While there certainly are common developmental events that can go awry to produce downstream effects years later, these can be disturbed in many different ways to produce the autistic phenotype. We know that there are many genes that predispose to autism. There are probably many environmental factors, such as toxins, infectious agents, and perhaps social experiences, which contribute as well. Moreover, each individual who developes the phenotype is unique; he or she becomes autistic through a different combination of genetic and environmental 'hits'. However, I think there are two fundamental principles of neuroscience that can help us clarify the autistic picture even further.

First, brain function is hierarchically organized. As we saw in the first two chapters, both the sensory and motor systems have lower level sensors and effectors on the one hand, and high level analyzers and controllers, on the other. The analysis of raw visual information becomes progressively more refined as it courses from the retina to the visual cortex. Motor commands are transformed from generalized goals to fine muscle movements as the neural code courses from the cord. Autism involves defects in both the sensory and motor realms. But they appear to lie at the **higher levels** of these realms. As a result, problems with fine motor control and visual acuity are not generally part of the autistic phenotype. Thus we should focus our search on the parts of the brain that mediate high level sensory-motor control, and cognition. One obvious area, as we saw in the second chapter, is the superior temporal cortex, where the dorsal and ventral visual streams, and the codes for perception and motion, merge. We know that autistic individuals seem to have functional and histological abnormalities in this region. Other possible sites of abnormalities include the thalamus, basal ganglia, cerebellum, and the PFC. Defects in brain areas subserving the 'lower level' functions, such as the primary somatosensory cortex, the retina, or the spinal cord, are much less likely to be directly involved in autism.

Second, the nervous system develops in a characteristic manner. The first stage involves the patterning of body axes and cell differentiation. This is followed by neuronal migration, the guidance of axons to their targets, and finally, the fine-tuning of synaptic connections. It is not yet clear where along the developmental pathway the autistic defect occurs. There are likely to be multiple points where it can occur.

Patterns

The human brain starts to develop from one end of a tiny streak on a flat sheet of cells. First, this 'neural plate' starts to fold into itself, forming a little tube. Different regions of this tube will form the brain and spinal cord. The genes that tell each region of the nervous system where it is on the anterior-posterior (head to tail) axis are remarkably conserved in all animals. These genes, in fact, were first discovered in the fruit fly.

On human chromosome 7p (the short arm opposite 7q, the site of the FOXP2 and WNT genes) there is a gene called HOXA1. It is likely to be one of the genes related to autistic susceptibility. HOXA1 is actually just one of a whole family of 39 hox genes clustered in four groups. Each group is located on a separate chromosome, with the HOXA group of 13 genes arranged on chromosome 7p15. The hox family of genes is very ancient, as is evident in their conserved sequence and organization across the entire animal kingdom from insects to humans. The ancestor of all hox genes originated some 500 million years ago, during the Cambrian geological period. The hox genes encode transcription factors that are expressed during the early stages of embryogenesis. Part of each gene contains a sequence of 180 nucleotides that is remarkably conserved across all the hox genes of all animals. This is called the 'homeobox'. The homeobox sequence codes for a special part of the hox protein that is designed to attach to certain stretches of DNA of other genes and affect their transcription.

The hox genes are turned on in the early embryo just after the neural tube (soon to become the spinal cord) fuses. The positions at which hox genes are activated follows a specific linear sequence: HOXA1 is anterior to HOXA2, which is anterior to HOXA3 and so on; the HOXB and HOXC genes follow a similar order. The truly amazing thing is that **the sequence of the hox genes along the chromosomal DNA corresponds to the order of the body segments that they help pattern**. In Drosophila, for instance, the first several hox genes pattern the head region, while the last hox genes pattern the abdomen of the creature. In mammals, as well, the first hox genes help form the upper vertebrae while the last ones form the tail. Anterior-posterior axis patterning in the animal kingdom is apparently so fundamental that its very order has been indelibly stamped in all of our DNA.

Several years ago, the mouse HOXA1 gene was successfully cloned. Scientists then inserted a defective 'knockout' copy into other mice. The resulting homozygous HOXA1 knockout mice had abnormal development of the brainstem, particularly involving the facial nucleus. They also exhibited abnormalities in the face, including external ear malformations and abnormal eye movements. The facial nucleus is a group of nerve cell bodies that project the fibers of the facial nerve, which controls the muscles of facial expression. HOXA1 normally directs the development of the brainstem level that contains the facial nucleus. Deleting or 'knocking out' HOXA1 literally causes a 'deletion' of that corresponding segment of brainstem.

Embryologist Patricia Rodier has found that brain samples from several autistic individuals exhibit a similar defect to the HOXA1 knockout mice [**Rodier, P (2000)'The early origins of autism' <u>Scientific American</u> 282(2) 56-63**]: the brainstem was shorter than normal, and lacked the layer that normally contains the facial nucleus. In addition, these patients, like many others with autism, apparently had facial and ear abnormalities and strabismus. Rodier reasoned that perhaps a defective HOXA1 was behind at least some cases of autism. She has found one variant of the HOXA1 gene that is more common in autistic than normal people, although this variant is not likely to be the cause of autism.

Childhood strabismus, or 'crossed eyes', is often due to a developmental problem with one of the cranial nerves or nuclei controlling the eye muscles. Intriguingly, many causes of strabismus seem to be associated with autism. Moebius syndrome, a genetic disorder caused by a defective HOXB1 gene, is associated with severe facial paralysis, strabismus, and autism. Children with Moebius syndrome can undergo surgery to restore some facial movements (such as smiling) and to correct the strabismus. If corrected early enough, the autism is reversed or prevented. This is unlike the more common form of autism, which is generally irreversible. There is some speculation that this may involve the 'mirror neurons' and the 'direct matching hypothesis' (chapter 3).

A related point is the finding that quite a number of children born to mothers who had taken the infamous morning-sickness drug thalidomide in the sixties had strabismus and external ear abnormalities (but not the more commonly associated limb malformations). About 5% of them were also autistic. This is 30 times greater than in the general population. It is known that the babies with the ear and eye problems but normal limbs were exposed to thalidomide around days 20-24 of gestation. This may indicate that thalidomide can suppress the normal action of the hox genes at this early stage of development, and this somehow unmasks the autistic phenotype some years later. Dr. Rodier is now investigating the role of thalidomide and other teratogens on the developmental genes.

How important are the hox genes in the pathogenesis of autism? Considering the modest allelic linkage to autism, and the similarity between brainstem abnormalities in HOXA1 knockout mice and some autistic individuals, there may be some role for the hox genes. But the linkage map analysis evidence is not very strong. In addition, most autistics do not have strabismus, let alone Moebius syndrome. A more likely explanation is that environmental factors, such as certain drugs, viral infections, or other prenatal insults, adversely affect the expression of certain varieties of hox genes leading to an increased susceptibility to autism.

Other environmental factors which have been implicated in 'priming' the autism susceptibility genes include prenatal maternal use of the antiepileptic drug, valproic acid, excessive alcohol use, congenital cytomegalovirus and rubella infections, and controversially, the childhood measles-mumps-rubella (MMR) vaccine. There is insufficient evidence to link any of these factors to autism. In the case of the MMR vaccine, there is quite a lot of evidence **against** a causal association. Unfortunately, this has not stopped many parents from unwisely refusing vaccination for their children. We clearly need to do more research before making any hasty recommendations.

Hox genes control the patterning of the anterior-posterior axis, a very early step in brain development. We know from studies like those with thalidomide-induced birth defects that at least some cases of autism do involve defects at this stage. Other cases of autism involving underdeveloped regions of the cerebellum or oversized brains are likely to be associated with defects at the later stage of cell growth and pruning. But many autistic brains don't have these features. Problems in earlier stages of development can certainly cause autism by disrupting subsequent development, but that does not help us pinpoint the etiology. What we need to do is concentrate on the 'least common denominator'. In this case, the latest developmental stage at which autism can occur.

The features that are common to the overwhelming majority of autism cases are the social, communication, and high order sensory-motor deficits. These deficits are likely to involve derangements in the later stages of development: axon guidance and synaptic fine-tuning.

Plasticity

There is growing evidence that autism is caused by defective neuronal organization in high order regions of the central nervous system. Neuroscientist Nancy Minshew [**(1996)'Brain mechanisms in autism: functional and structural abnormalities' Journal of Autism and Developmental Disorders** 26(2) 205-209] writes:

'Autism is now widely viewed as a developmental disorder of the brain involving neuronal organization. Several events in neuronal organization have been implicated, specifically the elaboration of dendritic and axonal ramifications, the establishment of synaptic contacts, and programmed cell death and selective elimination of neuronal processes. The resulting functional impairments involve cognitive processing, neocortical circuitry, and higher order abilities across domains. Collectively, these findings suggest that autism is a developmental disorder at the neural systems level of brain organization, resulting in generalized impairments in complex information-processing abilities subserved by these systems. The clinical syndrome is therefore most likely to be unified at the biologic level by a common dependence on a specialized type of neural connections or networks subserving higher order abilities.'

But precisely **which level** of organization is disrupted? Are the nerve cells in the wrong place? Are there too many or too few cells? Are the cells miswired? Are the wires providing the proper signal strength? These are open questions for researchers at the moment. There is some evidence for each of these themes. However, for several reasons, I believe that the main problem is at the final level of organization: **synaptic plasticity.**

The brain is not just a haphazard mass of neurons. Proper function depends on proper connectivity. The later stages of brain organization are defined by the growth and generation of axonal connections. The first step is called axon guidance, which refers to the process by which nerve cells in one part of the body find their targets in another part. This is a remarkable phenomenon. Imagine an upper motor neuron in the left motor cortex area representing the extensor muscles of the right thumb. It sends an axon branch deep into the white matter, through the brainstem, across the midline of the medulla, down the right spinal cord, to synapse on its precise and proper lower motor neuron, which, in turn, will sprout an axon destined to travel to its target, the motor endplate of the extensor digitorum profundus muscle which extends the right thumb many centimeters away. This axon is traveling with millions of other axons originating from other parts of the brain, yet it maintains its separate identity throughout the entire journey, protecting the integrity of the motor system's somatotopic organization. This happens with all the motor neurons of every developing embryo. The same thing happens in the sensory system. Sensory neurons in the skin or inner ear or retina sprout axon fibers that find their way through great distances to a precise location on the thalamus. There are, in fact, little 'maps' on the surface of the thalamus that correspond to contiguous areas of the skin, parts of the visual field, or changes in pitch, that are represented by the neurons whose fibers converge there. Second neurons from the thalamus maintain this precise connectivity and send axons to specific regions in the cortex where even more elaborate maps are found. How in the world do these axons know where to go?

The axon's leading edge is called a 'growth cone'. It has no tiny 'eyes', 'ears', or 'brains' of its own. It does have an army of specialized molecules on its surface that sense other chemicals in the environment. The growth cone slowly moves along (up to a millimeter a day), being attracted or repelled by the chemicals it encounters along the way. In this way, the axon can travel great distances and make precise connections. Scientists have now isolated many of the proteins involved in the axon guidance process.

Intriguingly, an international team of researchers has recently found evidence that mutations in the genes for some of these proteins (called cell adhesion molecules) are linked to autism. [**Jamain, S et al, 2003, 'Mutations of the X-linked genes encoding neuroligins NLGN3 and NLGN4 are associated with autism'** <u>**Nature Genetics**</u>]

◆   ◆   ◆

The next level of organization involves the strengthening and stabilization of specific synaptic connections. This is synaptic plasticity. Unlike the earlier developmental processes (pattern formation, cell differentiation and proliferation, and axon guidance), which are largely gene-driven, synaptic plasticity is the mechanism by which our brains respond to the environment. The genes we inherit from our ancestors specify biological order at a coarse level, but every individual's unique life experiences sculpt the fine detail. The trillions of precise connections between billions of neurons are constantly formed, pruned, and refined through continuous neural activity. The electro-chemical code of individual experiences, rather than the genetic code of DNA sequences, is the instruction directing synaptic plasticity.

To better understand how experience molds the brain, we will examine two unrelated phenomena: the development of ocular dominance columns, and the establishment of long-term memories. As I mentioned in chapter 3, visual stimuli are perceived by the retinal cells of the two eyes and passed on to the primary visual cortex (V1). Throughout this entire pathway, the activity from each eye is segregated into separate 'channels' all the way to the cells of V1. We know this because if we label the retinal ganglion cells of each eye in an experimental mammal (such as a cat or a ferret) with different colored dyes or 'tracers', they project to separate regions in V1. These subdivisions of the visual cortex are called 'ocular dominance columns' because they alternate with activity from each eye. Also recall from chapter 1, that these columns are subdivided into orientation specific and color specific areas to form 'hyper columns' throughout the visual cortex. The pioneering neuroscientists David Hubel and Torsten Wiesel discovered that the mammalian ocular dominance columns form after birth. Monkeys who were deprived of vision in one eye at birth (by having the eye sutured shut) failed to form ocular dominance columns. Instead, cells in the visual cortex could only respond to input from the open eye. In fact, that individual became permanently blind in the sutured eye after it was reopened. If an eye is closed at progressively later periods of postnatal development (such as two, three, or six weeks in the case of cats), the columns from the shut eye do develop to a progressively greater extent, although not to the same extent as those for the good eye. Finally, **adults** that had one eye shut, even for years, had no change in the size of their ocular dominance columns and were perfectly able to see after the eye was re-opened. There is, then, something molding the organization of cortical columns during a critical period in the first weeks of infancy.

It turns out that this process is driven by synchronized neural activity. Visual input activates the cells of the LGN, which send axons to the cortex, forming the cortical columns. Neurons in the cortex that receive synchronized activity from multiple neurons (such as several LGN cells sensing the same patch of white from a star in the American flag) will tend to form stronger connections with those LGN cells than to other LGN cells that do not fire at the same time. Neurons that fire together, wire together. But those cells that receive only unsynchronized activity during the critical period tend to lose connections or even die off. This is the essence of neural pruning: 'use them or lose them'.

Unlike the ocular dominance columns, which form in response to experienced visual activity, the eye-specific LGN layers are already well formed at birth. Since it's pretty dark in the uterus, the developing fetus is effectively blind. This seems to violate the principle of activity dependent plasticity. But, in a remarkable discovery, it was found that there are spontaneous synchronized 'waves' of retinal activity during fetal life. [**Meister M, et al (1991) 'Synchronous bursts of action potentials in ganglion cells of the developing mammalian retina' Science 252, 939-943**] The activity does not correspond to anything actually 'seen', and the fetus is certainly not conscious of them, but they exist, nonetheless. The waves themselves can be visualized using special calcium sensitive dyes on ferret retinal preparations. These neural waves from the two sightless eyes direct the proper segregation of the eye-specific layers of the LGN. The activity need not be conscious, or even externally derived, as long as it is synchronized.

However, the organization of the visual cortex, and visual consciousness itself, depends on visually experienced activity during the critical period. Experiments on fish raised in complete darkness found that the adults were effectively blind. Those raised in dark rooms punctuated by random flashes of strobe lights could see, but not very well. Only those fish raised in normal environments, seeing the plethora of objects they would normally encounter in their world, had fine-tuned visual maps in their brains. The crispness of neural connections depends on perceived differences in the environment. The smaller the group of cells that receive the same input and fire in synchrony, the finer the map in the visual cortex.

◆   ◆   ◆

In chapter 6, we discussed the different types of memory. The distinctions between short, intermediate, and long-term memories seem to have a molecular basis that involves synaptic plasticity at the level of a single synapse. The neuroscientist Eric Kandel has studied synaptic transmission in the sea slug, 'Aplysia', as a model for learning and memory in higher organisms. Using this little creature, he found that there are two separable processes governing short-term and long-term memory. An example of short-term memory in Aplysia is the ability to associate a noxious stimulus, such as an electric shock to the tail, with an otherwise harmless one, such as a gentle touch of the siphon (blowhole). After a few trials pairing the two stimuli, the creature learns to respond vigorously even when it receives only the conditioned stimulus (light touch) by itself. Furthermore, it will 'remember' to do this for several minutes before forgetting, and once reconditioned, it will 'learn' faster the second time. In the case of long-term memory, the animal must receive many trials of paired stimuli (the learning phase), spread out over several sessions (the consolidation phase). After this, they remember to associate stimulus 'A' with response 'B' for many days. These two processes involve fundamentally different mechanisms at the molecular level, and they appear to be universal to learning and memory in all higher animals.

We will now make a brief detour to examine normal synaptic transmission. Signals are transmitted from one nerve cell to another across the thin space between presynaptic axon and postsynaptic dendrite. This is called the 'synaptic cleft'. The cleft is only about 5 nanometers wide (there are a thousand nanometers in one micron, and a thousand microns in one millimeter), but the electrical action potential in one active nerve cell cannot cross over to stimulate the next cell across the gap. When the electrical activity reaches the presynaptic end of its axon, it triggers the release of stored neurotransmitter chemicals, usually glutamate or acetylcholine, but sometimes GABA, serotonin, dopamine, norepinephrine, histamine, or a whole list of others, depending on the cell type. The neurotransmitter then diffuses across the cleft and binds to its receptor on the dendrite of the postsynaptic cell. Each neurotransmitter has at least one, and often several separate receptors. Receptors come in one of two main forms: 'ionotropic' and 'metabotropic'. Ionotropic receptors, which we will be mainly concerned with, are simply membrane channels that allow ions of sodium, potassium, calcium, or all three to cross the cell membrane when activated depolarizing the target cell and inducing it to fire an action potential. Metabotropic receptors are much more complex creatures with seven membrane spanning regions and linkages to special enzymes that generate so-called 'second-messengers'. Second messengers are small intracellular molecules that transmit signals to other proteins and ultimately to transcription factors, which turn other genes on and off.

Certain neurotransmitters, such as serotonin and dopamine, are involved in modulating the activity of target neurons. Cells that release these chemicals often originate in subcortical or brainstem areas and synapse on cortical cells. When cortical neurons are simultaneously activated by these modulatory signals as well as by their 'normal' input, the input signal is imbibed with a special memory or valence. This sort of modulation is likely to be important in the way the subcortical regions such as the amygdala, hippocampus, basal ganglia, and the brainstem reticular formation interact with the higher cerebral cortex.

In terms of short-term associative learning, two separate input signals (representing the two items to be associated: touch=shock, crime=punishment, smile=happiness) that simultaneously activate a target cell, will depolarize that cell more than a single stimulus would. Low levels of glutamate released from the presynaptic terminal of a single cell can activate only the first and simpler of the two glutamate receptors, called the nonNMDA ionotropic receptor. It simply allows the initial depolarization of the postsynaptic cell by letting sodium and potassium ions pass through the cell membrane. If this depolarization is accompanied by an additional influx of glutamate from a second presynaptic cell, this glutamate can now bind to the second receptor, called the NMDA ionotropic receptor. The NMDA receptors open when multiple activating signals converge on the postsynaptic cell. The NMDA receptor facilitates a longer lasting depolarization by opening calcium channels. When calcium then enters the dendrite, certain enzymes are activated that open more nonNMDA receptors. In addition, 'retrograde' chemical signals are generated that enhance synaptic transmission from the presynaptic axon, further strengthening the synaptic connection. This is how short-term learning and memory is believed to occur at the molecular level. The key is 'heterosynaptic' or 'Hebbian' plasticity: multiple simultaneous input signals (via multiple synapses) upregulate and change the responsive properties of the postsynaptic cell.

Long-term memory requires the further step of gene transcription, and the physical growth of new dendritic branches. Similar to the case of short-term memory, the nonNMDA and NMDA glutamate receptors are activated as before. The difference is that now, the input is not just synchronized activity, but **repeated trains of spaced out synchronized activity**. This recruits more NMDA receptors, facilitating even more calcium influx. The calcium influx activates second messenger systems, which, in turn, start enzyme cascades culminating in the activation of certain genes involved in synaptogenesis. This synaptogenesis, the sprouting of new dendritic arbors, and the physical remodeling of the brain consolidates short term learning into long-term memories.

In the developing nervous system, a process similar to memory is believed to consolidate new synaptic connections, such as those in the visual cortex, through synchronized activity. The role of NMDA receptors in activity-dependent synaptic plasticity has been demonstrated in a set of elegant experiments conducted by Martha Constantine-Paton and her colleagues on the visual system of the frog [**Constantine-Paton, M et al (1990) 'Patterned activity, synaptic convergence, and the NMDA receptor in developing visual pathways'** <u>Annu Review of Neuroscience</u> 13, 129-154]. Unlike cats and humans, the frog normally keeps visual input from its two eyes in separate channels all the way to the visual cortex. Since there is no competition from the two eyes for access to cortical representation, no ocular dominance columns develop. However, if a **third eye** is experimentally implanted into the frog's brain, ocular dominance columns form! This indicates that it is the synchronous input (coming from the extra eye and one of the two original eyes, in this case) that activates NMDA receptors and downstream second messenger systems and strengthens some synaptic connections relative to those that are not synchronously activated. Moreover, when NMDA receptors were artificially over stimulated in these frogs, the columns 'sharpened' in resolution. When NMDA receptors were blocked, the columns became fuzzy or disappeared entirely! It is as if the activity of NMDA receptors controls the signal/noise ratio or the 'gain' of synaptic plasticity.

◆   ◆   ◆

In brain development, we know that environmental input is necessary for proper synaptic remodeling. We know that synchronized activity is needed to activate the NMDA receptors to strengthen synaptic connections. But what are the signaling molecules involved in these connections? What tells the presynaptic axon that it is now in the vicinity of the proper dendrite? What is the chemical nature of the retrograde signal telling the axon to proliferate or retract? One intriguing idea is that the major histocompatibility complex (MHC) molecules, which are important in immune system signaling, may also be involved in neural plasticity. Carla Shatz has found evidence that knockout mice deficient in MHC components suffer abnormal synaptic connectivity between retinal neurons and the LGN [**Shatz, C (2000) 'Functional requirement for class I MHC in CNS development and plasticity'** Science **290, 2155-2158**]. In addition, these mice demonstrate abnormal LTP within the hippocampus, the site of memory consolidation.

Other molecules potentially important in synaptic plasticity include, of course, the motley crew of neurotransmitters and their receptors, as well as various chemicals that activate nerve growth and proliferation, called 'neurotropic factors'. If autism were a fundamental defect in synaptic plasticity, it would involve defective proteins at this level. We should look for mutant genes or abnormal proteins involved in neurotransmission and synaptic stabilization. There is as yet little evidence that MHC molecules are involved in autism. But there have been some suggestive finds elsewhere.

◆   ◆   ◆

Earlier, we reviewed the results of genetic linkage studies that have narrowed down potential autism causing regions of the human genome. Several of these studies have focused on candidate neurotransmitter system genes. There is some evidence that serotonin, dopamine, GABA, or glutamate transmission may be dysregulated in autistic individuals.

Serotonin and dopamine are specialized modulatory neurotransmitters that facilitate cortical-subcortical communication. Serotonin is especially important in the mediation of general arousal and mood. Low serotonin levels have been implicated in depression. There is evidence that serotonin also acts as a neurotrophic growth factor during early development, possibly acting to stabilize synapses or promote synaptogenesis. Some autistic subjects have been found to have high levels of serotonin in their blood, but are less responsive to its effects, perhaps due to down-regulation of serotonin receptors. Selective serotonin reuptake inhibitors (SSRIs) such as prozac and paxil, which are useful in the treatment of depression and anxiety by increasing serotonin levels, are also known to improve certain autistic symptoms. These findings have led some researchers to test the linkages between the genes involved in serotonin metabolism, and autistic families. The results have been mixed, and there is still little convincing evidence for any genetic linkage. The same has been true for candidate genes involved in dopamine and GABA transmitter systems. The most promising lead is a possible linkage with a glutamate receptor.

GluR6

A group of French biologists led by Stephane Jamain has found significant evidence of linkage between a gene on chromosome 6q21 and autism. The gene codes for a glutamate receptor, more specifically, the sixth subunit of the 'kainate ionotropic receptor', or GluR6. [**Jamain, S et al (2002) 'Linkage and association of the glutamate receptor 6 gene with autism'** Molecular Psychiatry 7, **302-310**]. GluR6 is part of the nonNMDA ionotropic glutamate receptor family. These receptors are expressed during embryonic brain development and may, along with the other glutamate receptors, play an important role in activity dependent synaptic plasticity. Jamain and his colleagues isolated and sequenced the GluR6 gene in autistic individuals. They found that 8% of their subjects had a mutated version of the gene (compared with 4% in normal controls) which changes the amino acid sequence of the protein in a possibly important way. The defective protein subunit may not be targeted to its proper location at the synaptic membrane, and thus hinders the function of the receptor during a critical period of brain development. Another interesting finding was that in the autistic subjects, this particular GluR6 allele was inherited from the mother. Like the genes involved in Prader Willi/Angelman syndromes, and the putative X-linked autism susceptibility locus proposed by David Skuse, the GluR6 gene appears to be imprinted.

These findings do not mean that a defective variant of GluR6 inherited from the mother causes autism. Four percent of nonautistic people seem to have it too. But it may, along with the FOXP2 and HOXA1 genes, be an important autism susceptibility gene. First, there is a fair degree of linkage to autism here. Second, the glutamate receptors, both NMDA and nonNMDA, play crucially important roles in synaptic plasticity. The GluR6-autism link may be a key to autism's source.

Knockout Mice

If GluR6 is an autism susceptibility gene, we would like to know where and when in the developing embryo and growing child it is expressed, and how an individual deficient in this protein would behave. It is rather difficult (as well as unethical) to try to answer these questions using human subjects. But a logical solution is to produce mouse knockouts at the GluR6 gene locus.

In short, here is how to make a knockout mouse. First, we start with a copy of the cloned mouse gene of interest, in this case, GluR6. [I explained how to get to this stage earlier in the chapter]. Mice and humans share much DNA homology. Using standard recombinant DNA technology, we insert a gene that confers resistance to certain antibiotics into the middle of the GluR6 gene. This has the dual effect of inactivating ('knocking out') the gene of interest, and also allowing only those cells which happen to take up this synthetically created piece of DNA to survive antibiotic treatment. Second, this 'recombinant' construct is then inserted into mice embryonic stem (ES) cells cultured from dark colored mice. Rarely, this construct will integrate into its correct (homologous) site on the mouse cell's DNA. This is called a 'targeted insertion'. Those mouse ES cells with the targeted knockout insertion can then be selected out using the antibiotic. Third, these cells are then inserted into otherwise healthy four-day-old light colored mouse embryos, called blastocysts, which are hollow balls of cells. The blastocysts are then reimplanted into a surrogate mother. The resulting offspring mice will have a mixture of cells descended from the dark colored knockout ES cells, and cells of the light colored embryo into which the ES cells were inserted. It will have light and dark stripes of fur since the animal is a 'mosaic' of cells from two different colored strains. Some, but not all, of these mosaics will have the knockout gene in their germline cells and can be used to produce offspring with one copy of the gene knocked out. Finally, to produce homozygous knockout mice mutant for the GluR6, we cross these mice to each other, in hopes that two mice with germline knockout mutations will meet.

Once the GluR6 knockouts have been generated, we can study the animals for behavioral, anatomical, and biochemical defects. If these creatures display defects in socialization or motor control, or evidence excessive repetitive behavior, it would support the fact that we have created an 'autistic' animal model. If this phenotype is indeed found, the next step would be to localize the function of the protein temporally and spatially in the developing mouse. Scientists have perfected ingenious techniques to do this. They are called 'conditional knockouts'. They cross two different strains of genetically engineered mice, the first has a form of the gene of interest that can be deleted when treated with a specific enzyme. The second is a 'transgenic mouse' with a tissue specific promoter spliced to a gene for this enzyme. Thus the hybrid knockout mice will have a defective copy of the gene only in those tissues where the promoter is active, such as the hippocampus, or basal ganglia, or amygdala, but not in the heart, or skin, or eye. Second, scientists can regulate when the gene is to be shut down during development by injecting a drug that activates the enzyme that shuts down the gene. By shutting down the gene at different periods of development, we can see what behavioral and physiological abnormalities result in these conditional knockouts.

An example is a knockout of the NMDA receptor specifically in the hippocampus. Scientists have found that mice have topographically organized maps of three-dimensional space coded within specialized neurons (called 'place cells') in their hippocampi. Recall from chapter 3 that primates seem to have similar space maps, in both egocentric and object-centric coordinates, in the right superior temporal cortex. Different place cells represent a different aspect of personal space, which the mouse can remember. NMDA receptor knockout mice have been created in which a subunit of the NMDA receptor is deleted out of certain cells of **only the hippocampus**. These so-called 'conditional knockout' animals were found to have less precise or 'fuzzy' mental maps of their personal space; individual place cells were not as fine-tuned to their normal representation.

Finally, we can use knockout mice to find downstream effects of the initial mutation. Recall that dysfunctional synaptic plasticity can result in other genes being abnormally turned off (or on). We can isolate total mRNA from these knockout animals, label it with a fluorescent tag, and pour the labeled mRNA over glass wafers (called 'gene chips' or 'DNA microarrays') embedded with thousands of single stranded DNA copies of normal genes arranged in a two dimensional array. If a specific mRNA is present in the tissue extract, it binds to its corresponding DNA on the glass chip and lights up. If not, it stays dark. A computer scans the chip and prints out the presence and intensity of each gene expression signal. Cutting edge technologies such as gene chips allow us to tell where, when, and how much of thousands genes are expressed in development.

Fuzzy Dice

I think that defects in GluR6 and/or some other genes for neurotransmitter system proteins or synaptic plasticity factors are likely to turn out to be the genetic cause for autism. The reason is that these genes are so instrumental in cell-cell interaction and communication at the synaptic and cellular levels of neurodevelopment. And I believe that autism is basically a disease of synaptic and cellular neurodevelopment which simply has lots of downstream ramifications at the anatomical and behavioral levels. These genes specify the level of fine-tuning the nerve connections achieve when exposed to synchronized activity either from elsewhere within the brain or outside in the world. Without them, the brain will still form, all the right cells will differentiate and migrate to the right place, and even most of the axons will find their proper targets. The autistic individual can still see, think, move, cry, and remember. But not in a fine-tuned way.

It has been discovered that brains of autistic individuals often demonstrate loss of Purkinje cells in the cerebellum and diminished dendritic arborization in the limbic system. It is also known that the addition of glutamate, nerve growth factor, or other neurotropic chemicals increases the dendritic arborization of the Purkinje cells. It is possible that the absence of the GluR6 or other related proteins in autistic children curtails the development and plasticity of specific parts of the CNS such as the limbic structures, the thalamus, or perhaps cortical/sub-cortical connections in general.

I propose that a diffuse disconnection between the cortex and subcortex may generate the autism type pathology. One major domain that would be affected involves the limbic circuits: amygdala-temporal lobe, and amygdala-PFC, producing the emotional pathology of autism—and perhaps a defective TOM. Another domain involves the subcortical motor circuits: basal ganglia-cortex, and cerebellum-cortex, producing the repetitive behavior of autism. Finally, a third domain involves the thalamo-cortical circuit, which disrupts attentional control, resulting in executive dysfunction and perhaps loss of central coherence. What these circuits have in common are billions and trillions of fine tuned connections between the cerebral cortex, where the contents of consciousness reside, and the subcortical organs, where the process of consciousness automatically takes shape. If the fine-tuning is lost, there is a loss of gain, a loss of signal to noise resulting not in a loss or altered state of consciousness, but rather a 'fuzzy consciousness'. This has something in common with the case of the fuzzy ocular dominance columns in the three-eyed frogs with blocked NMDA receptors, or the case of the fuzzy hippocampal place cells in the NMDA knockout mice. All are consequences of defective synaptic plasticity at the molecular level.

◆   ◆   ◆

Attracting women is a great mystery to me. Will someone tell me how it's done? I've tried being forward and assertive, I've tried the quiet and sensitive schtick, I've experimented with being different and original, I've worked very hard at being normal and conventional, and I've even settled on just 'being myself'. Nothing seems to work. Is it because I have glasses? Is it because I'm short? Is it because I can't tell jokes? Is it because I have no friends? I see other guys getting lucky at bars and parties. I've seen insensitive cads with gorgeous blondes; I've seen ignorant jocks win intelligent hearts. For a long time all I seemed to attract was ridicule and gay men.

Growing up, my attitude to sex fluctuated between skeptical indifference and jealous hostility. I honestly couldn't understand why all the Beatles songs (at least the early ones) were about girls, love, and handholding. Sex was something funny and dirty that all my classmates incessantly talked about. By junior high school, the most popular boys in class, who also happened to be the best at sports and at teasing kids like me, started making out with the most popular girls. I saw them kissing by the lockers, behind the gym, in the back on the school bus. I tried to rise above it, thinking sexuality was irrational, and somehow 'beneath' me. It certainly didn't rate as high as schoolwork. But I had a gnawing suspicion that perhaps I was missing out on something important.

In high school, it wasn't just the popular kids that got to make out. The 'bad kids' and even some of the regular kids got into it as well. Needless to say, I was neither bad nor regular. To make out, one needs to 'hang out': at parties, friends' houses, after school. To hang out, you need to have friends to hang out with (or at least access to a car). I had neither. It was a vicious cycle of social rejection leading to psychological dejection leading to still more rejection.

In college, the pattern simply got worse. The more antisocial I was, the harder it was to make friends; the fewer friends I had, the harder it was to make new friends, or even to believe that I deserved friends. Sex was not only an unlikely prospect, it wasn't even a priority. Happiness was achieving academically, or watching science documentaries in my dorm room or browsing the bookstores in Harvard Square. I did enjoy the presence of a few 'kindred souls' with whom I had late night talk sessions, but they certainly didn't make it any easier to find female companionship.

After college, I decided to study British history and philosophy at Oxford. As an undergraduate, I had developed an admiration for all things English: movies and documentaries, accents, history, architecture. A naive goal of mine was to travel to all the territories of the former British Empire and to somehow win the love and affection of the English upper class. But to my dismay when I finally saw Oxford, the distant dreamy spires in the post cards and travel books were actually quite grimy up close. The city was crowded with tour buses and loud Americans. Some of the students may have been smart and posh, but most were like others back home: out for a pint and a good party. My anglophilia and aristocratic pretensions were simply ridiculous in the context of my general social ineptitude, which transcended the bounds of class, culture, and ethnic origin.

By the time I started medical school, my outlook became more pragmatic, my politics more conservative. But as always, there were too many other things to do, too many new things to learn. Sex and romance were not among them. In the summer of my 25th birthday, I did meet a young woman on a three-week youth camping trip to the Canadian Rockies. Sarah was a comely 20-year-old nursing student from Bexhill on the East Sussex coast. She was, to my eyes, the ideal English Rose: tall, graceful, aristocratic, blond; a doctor's daughter from the Home Counties who spoke the Queen's English. I had finally found the woman of my imagination.

Alas, it was not to be. It would never be enough to admire her feminine charms from afar. But I didn't know what to do. It took a week just to get up the courage to talk to her (I was terrified she might find me too forward); she probably thought that I wasn't interested. The second week, I did everything I could to be near her. After some awkward walks, talks, and misdelivered jokes amid spectacular glaciers and alpine lakes, the climax of our relationship was a drunken kiss on the last night. It took me two years and a trip to England (where she showed no interest in reviving our relationship) before I was finally over her.

The target of my next abortive attempt at courtship was Nora from Donegal, whom I met in a New York hospital in the spring of 1996. Some months after that, my bar crawling paid off: I met another feisty young nurse, the Irish woman from Belfast who became my first real sexual experience. At 28, I was still learning to connect.

8

Chink

Without the impulse of the individual, the community stagnates.
Without the sympathy of the community, the impulse dies away.

—William James

Hapsburg Vienna, February 18, 1906. Here was the absolute crown jewel of European Jewish high culture glittering brilliant in its Indian summer. The galleries, salons, theaters, ballrooms, and opera houses around the Ringstrasse buzzed with the creative energy of Gustav Klimt, Oskar Kokoschka, Arthur Schnitzler, Gustav Mahler, and Sigmund Freud. A 32 year old composer named Arnold Schönberg was about to invent the 12 tone scale, an 18 year old undergraduate named Erwin Schrödinger had yet to invent quantum mechanics, and a 17 year old named Adolph Hitler was dreaming of becoming a great painter. Into this world was born a certain Hans Asperger. No one realized what momentous influence he would eventually have on a young Korean-American boy a century and an ocean away.

Nazi Vienna, 1944. Hans Asperger, just back from his duties as Wermacht captain in German-occupied Croatia, first described a new type of 'autistic psychopathy'. His scientific papers (written in German) made no international impact at the time as Europe was in flames. In the east, Soviet tanks were demolishing Hitler's empire; to the west, the Anglo-Americans were preparing to invade the beaches of Normandy. Within a year, Russian troops and Nazi fanatics would engage in point-blank artillery duels down the center of once-elegant Viennese boulevards. If Asperger's findings had been published prior to the expulsion of Jewish intellectuals, things might have turned out differently. Austria and Germany, after all, were at the cutting edge of art and science before the Nazi takeover. In the end, news of Asperger's curious syndrome had to wait until 1981 to reach the outside world. That year, a young British psychologist named Lorna Wing translated Asperger's manuscripts into English. The rest is history.

That same tumultuous year of 1944, my father was born in a small town in the Korean province of the Japanese Empire. His father was an intelligent but unsuccessful man from peasant stock. He worked for a time as a private tutor to the woman who would later become my paternal grandmother. By most accounts, he was obstinate and domineering, traits he would pass on to my father. My father's mother suffered a debilitating stroke before I was born, so my knowledge of her personality is rather limited. However, I do know something of her character flaws. She was self-centered and inflexible. She wanted to study in Japan, but her father forbade it. In an impulsive rage, my grandmother flung all her books into the fire and vowed never to return to school. She had few friends and even fewer suitors. It was fortunate that she met my grandfather.

They had three daughters and a son, who would later become my father. The eldest daughter became a pharmacist and immigrated to the United States (along with just about everyone else in my extended family) in the early 1970s. She and her husband subsequently settled in Atlanta, and had two daughters, one of whom is a physician, and the other, an engineer. A second aunt became a biochemist, settled in Kentucky and married an American colleague. Another aunt is a psychiatrist outside Detroit. Now in her sixties and unmarried, she shares many of her mother's asocial traits: a labile personality, rigid behavior, and temper tan trums.

My father claims he overcame great burdens of poverty. I recall his stories about having to walk miles to school each day with tattered old books in hand and broken shoes on his feet. He was academically successful, but had difficulty socializing with the other boys. He was often teased and bullied. My father was an introvert during his years in Yonsei medical college. One of his passions at the time was German philosophy. It was in medical school that he met his first girlfriend, the young music student who would become my mother.

My mother was born in the winter of 1946 in what is now North Korea. At the time the land was part of a fledgling Korean state occupied by Soviet troops. Communists had infiltrated all levels of leadership following the Japanese collapse the previous summer. My mother's paternal grandparents had become relatively wealthy during the 1930s by collaborating with the Japanese government. My great grandfather owned and managed a textile factory near Harbin, deep in occupied Manchuria, and used the profits to secure a rather comfortable life for his family. But as the Sino-Japanese conflict expanded into a global war against the Anglo-Americans, their way of life was threatened. My mother's father, then a student of European languages at Tokyo University, was recalled home with his young bride. He would tell us about the vapor trails of B-29 bombers flying high overhead, followed by evacuations to the air raid shelters. By the time my mother and her older brother were born, nuclear bombs had devastated Japanese cities. The terrible 35 year occupation was over. Thirty million jubilant Koreans were suddenly free. But these were exceedingly dangerous days for well to do Koreans like my mother's family. First Soviet and then Red Chinese troops pushed into Manchuria and then advanced down to the 38th parallel, spreading communist propaganda to the uneducated masses. Many were converted by promises of a socialist revolution. But in a communist country, everything is free, except you. Private property was looted. Intellectuals, professionals, and anyone thought to have connections with the Japanese were ruthlessly hunted down by left-wing death squads. My mother's family abandoned its house and fled on a hazardous sea voyage to U.S. occupied Seoul to the south.

Four years later, on June 25, 1950, communists from the Soviet zone invaded the unprepared South, and promptly captured Seoul. My mother's family had once again become refugees. This time they fled to the port of Pusan, at the southern tip of the peninsula. There was nowhere else to hide, and nothing else to do but hope and wait for liberation by the Americans. Fortunately, UN troops were able to hold off communist tanks outside the city just long enough for General Douglas MacArthur to arrive. U.S. marines landed in Inchon, recaptured Seoul, cut off and annihilated the retreating North Korean army, and advanced up the peninsula all the way to the Chinese border. Many civilians (but fortunately not my grandparents) thought the war was over, and started to return home. Then that winter, a million screaming Chinese communists crossed the frozen Yalu River and slaughtered thousands of American boys on the other side. Countless starving and exhausted Korean refugees once again fled southward. MacArthur wanted to punish the Chinese with a nuclear counter-attack. Fearful of inciting Stalin into a third world war, President Truman fired MacArthur. The struggle dragged on in a bloody stalemate for another two years. Seoul changed hands four times in the course of the conflict.

In the summer of 1953, the fighting ended and the family moved back to the ruins of Seoul. Everything had changed. Three million Koreans (10% of the population), a million Chinese, and 55,000 Americans had lost their lives. Millions more were maimed, orphaned, or made homeless. One in three Koreans was forever lost inside the shameful black hole that is North Korea. The Demilitarized Zone still stands today, a gaping wound arbitrarily separating a people in two; it is a glaring and enduring testament to the failure of the Korean nation and the communist ideal. Like everyone else, my great grandfather had to start over. He never regained his wealth, but he continued to maintain an elitist bearing. It was all the more tragic in view of his diminished circumstances.

◆   ◆   ◆

I remember my father as being at times strict and harsh and at other times generous and playful. His wild mood swings were unpredictable. When in a bad mood, he would find an excuse to yell at, threaten, or humiliate my mother. My father also has other peculiarities. Once he becomes involved in a topic he finds interesting, he starts a rambling monologue, often losing the point and drifting off into irrelevant tangents. He talks at you, rather than with you. My father is also quite rigid in his opinions, but not necessarily in a consistent way. One day he may say that interracial dating is wrong or that the Democrats should run the country. The next day, he may insist that mixed marriages are fine, or that the Republicans are the better party. Alarmingly, I often notice these traits in myself.

Some of my most unpleasant memories are of my father 'teaching' me math. He believed (correctly) that the public schools were not doing a very rigorous job of it, so he decided to have a go at it himself. He would write out lists of algebra problems on a blackboard at home, quickly solve one or two in his own way, then leave me alone to figure out the rest. He would threaten me with physical punishment if I got it wrong. I can vividly recall sitting there sobbing, totally confused and terrified by all those obscure signs and numbers. Sure enough, he would return after a few minutes (which seemed like hours of torture) and, seeing me clueless, start pummeling me and yelling at my stupidity. But I could no longer concentrate, nor even see the blackboard through the torrent of tears streaming out of my eyes. I remember the salty taste of the viscous snot dripping down my throat. Sometimes I became dizzy, and felt myself leaving my body and floating above it, suddenly immune to his slaps and blows.

The worst memories are those of my father and his father ganging together and beating up my mother. One horrific night, my grandfather chased my mother out into the yard and into a neighbor's house with a kitchen knife in hand. He relented only when our neighbor threatened to call the police. I saw all of this through the terrified eyes of a helpless eight year old. It's remarkable that my mother didn't leave him. When I ask her why, she tells me she did it for the kids. I think it would have been very daunting for her to care for the two of us on her own, with little English and no job. I cried every time I saw the black and blue splotches on my mother's arms and legs after her beatings. I cried with pity for my mother. I cried with rage at my father. But when I think about it now, I think I cried mostly for myself.

Part I: Talk

To neurotypical ears, the language of autism seems very odd, indeed. Many autistics simply don't speak much at all. When they do talk, it is often limited to disjointed words or catch phrases repeated over and over at inappropriate times and places. We think of the Dustin Hoffman character, Raymond Babbitt from <u>Rainman</u>: '…of course, I'm a very good driver, a very good driver, I'm a very good driver, a very good driver…'. Autistics can be quiet when content, but noisy when agitated. The persistence of repetition is proportional to the degree of distress. As a tormented teenager, I would frequently vent my anger and frustration through stereotyped tantrums: '…the round thing goes around and around! The round thing goes around and around! The round thing goes around and around!' (either in English or Korean). Sometimes, I added a few expletives into the mix: '…the round fucking thing goes around and around! FUCK! The round fucking thing goes around and around! FUCK! The round fucking thing goes around and around! FUCK!…' I would usually have to repeat this exactly five or seven times (or multiples of five or seven, depending on my degree of distress), don't ask why. Sometimes five was holier than seven, but other times seven was holier than five. I was not trying to convey any information, but these outbursts simply and inexplicably made me feel calmer.

Autistics are known to be unconditionally honest, exceedingly gullible to lies and deceit, impervious to sarcasm, irony, and subtle humor, bad at fabricating and appreciating creative fiction, and prone to engage in endless monologues about things like UFOs, world capitals, sports statistics, and train schedules. What underlies these very odd aspects of communication? Is it some aspect of cultural learning or grammatical competence that autistic children miss at some crucial stage of development? Or rather is it something more fundamental, involving the patterns of thoughts, the circuitry of the brain, or the functions of genes? Is it one big defect, or lots of little ones acting together at multiple levels that set off an 'autistic cascade'? In this chapter, I will attempt to answer some of these questions by exploring the way that neurotypicals communicate and the defects in autistic communication.

Universal Grammar

Perhaps more than any other attribute, it is the ability to communicate via language (spoken, written, sign, etc) that distinguishes humans from all other animals. We don't know how language evolved since body language and voices don't fossilize. Written records take us back several thousand years, but speech and language were already well established by then. The psychologist Michael Corballis proposes that spoken language developed from a form of sign language some 100,000 to 200,000 years ago, corresponding to the last great human migration out of Africa [**Corballis, M, <u>From Hand to Mouth: the Origins of Language,</u> 2002**]. But his theory assumes that language (in the form of symbolic gestures) already existed long before we started to talk. We should start by examining the behavior of the great mass of life on earth that that has no language at all, non-human species.

All living things 'communicate' in the broadest sense of the term. Single-cell organisms secrete chemical signals to attract, repel or instruct other cells to grow, move, or produce more of this or that substance. In multicellular organisms, different tissues become specialized to communicate information for mutual benefit. The cells of the immune system provide data on the identity of foreign and self-produced substances. The cells of the nervous system provide information on the electromagnetic, mechanical, thermal, and acoustic properties of the immediate and remote environment. The cells of the endocrine system provide signals controlling metabolic functions of the entire organism.

Communication between individuals in the animal kingdom has evolved amazing complexity. Bumble bee dances, birdcalls, and whale songs all involve intricate coordination of muscles to produce stereotyped informative messages for the benefit of others. Through them, other bees find the source of food, birds defend their territory, and orphan orcas find their pods thousands of miles away. Some animals have even mastered symbolic representation. The vervet monkey is said to have at least three different calls in response to seeing a 'snake', a 'leopard', or an 'eagle'.

What makes human communication different is first, the aspect of attributing meaningful references to otherwise arbitrary physical acts (a flick of the finger, a hum, a click, a high-pitched shriek). Very few animals are capable of this. And second, the recombination of the meaningful units (words and word stems) in a linear sequence in time (phrases and sentences) to depict thoughts, stories, and desires. This double combinatorial system of sounds/gestures into meaning, and then meaning into stories is known as the 'duality of patterning'. As any student of elementary math knows, combinatorial systems generate an endless variety of possibilities from a limited number of building blocks. A finite number of hand and facial gestures, sounds, or letters of the alphabet can be used to generate a very large number of possible words, each with a slightly different meaning (semantics). Additionally, a limited number of words (roughly 20,000 per language, but over 100,000 for Standard English, not counting further hundreds of thousands of scientific and technical terms) can be used to generate a near infinite number of possible sentences, each expressing a different story. It is the second type of patterning, from meaningful units (words) to sentences, that is uniquely human. This is human grammar.

◆   ◆   ◆

Grammar, or syntax, is the way units of meaning are combined so as to become intelligible for those with whom one is communicating. Contrary to what one may believe, especially after struggling through language and grammar classes in grade school, the basic rules of grammar are not the product of some arbitrary cultural convention. They are hard wired into every normal child's brain. Children are not born talking, but by age two, they start constructing single words out of meaningless babble. By two and a half, they start combining words into simple sentences. By age three, their language development literally explodes into a non-stop riot of exclamations, exhortations, inquiries, and imperatives, many of them, quite unintentionally, profound, humorous, or both. The mother tongue and native culture may be radically different, but the unfolding of the developmental timetable is almost constant for every child on earth.

There is a commonality, not only to the childhood acquisition of language, but also within language itself. The basic components of the English sentence: subject, verb, object (or subject, object, verb in most other languages including Korean) are used by all children in all languages and cultures whether or not they learn them in school. Those often annoying language specific details such as gender, irregular verbs, and odd plural forms must, of course, be learned, along with the vast lexicon (vocabulary). This is what parents, teachers, and television are for. But 'basic grammar' does seem to be a universal instinct set to unfold in a specific way at a specific time.

Evolution of Talk

If there is a universal grammar or syntax independent of learning, culture, and one's particular language, how and when did it come about? Let's start with a few observations. First, no animals, not even dolphins, chimpanzees, or parrots, have any natural grammar. They are undoubtedly clever, emotional creatures with good memories and the ability to learn faces and signs, and perhaps solve a few simple problems involving symbolic logic. But there is really no evidence that they use a combinatorial system of syntax. Even with painstaking training, chimps are only able to master a few rules of grammar. It is likely that they learn them by memorization of particular examples rather than by conceptual generalization. Since chimps are our closest surviving genetic and evolutionary cousins, we can be fairly certain that grammar evolved sometime after our common ancestral lines diverged about seven million years ago.

Second, grammar is not limited to speech. Deaf people learn to communicate in 'grammatical' sign language. Deaf children raised in the absence of formal sign language, but in proximity to other deaf peers, spontaneously 'invent' their own gestural language. Such languages have their own syntactic organization, somewhat akin to the unconventional grammar of Creole languages, which develop from the talk of illiterate children of polyglot 'pidgin' speaking parents. Thus, grammar is not limited to speaking people with a literate background.

Then there are those people without grammar. Those born with the SLI (Specific Language Impairment) disorder discussed in chapter 7 are otherwise intelligent people born with a defective so-called 'grammar gene', the FOXP2 gene on chromosome 7. People suffering strokes to the left prefrontal region known as Broca's area usually lose (for various periods of time) the ability to speak and write grammatically, although their thoughts may be (largely) unimpaired. Finally, there are reports of unfortunate children who for one reason or another were socially isolated for the first few years of life and grow up to be partially or sometimes even totally incapable of grammatical speech.

These observations indicate that syntactic thought is independent of speech (multimodal), localized to fairly specific parts of the brain (modular), at least partly genetic, and relatively recently evolved (within the last seven million years). Aside from this, not too much is definitively known. A very interesting finding is that the monkey homologue of Broca's area (the presumed grammar center of the human brain) and the corresponding area on the right side contain the mirror neurons discussed in chapter 2. These cells fire when the monkey moves its own body and when it observes other monkeys making the same movement. It is possible that these brain regions first evolved to enable monkeys to recognize and understand the intentions of others via the observation of their movements in space. Then, they were later 'recruited' through evolutionary history to allow humans to understand and represent the intentions of others based on otherwise arbitrary temporal sequences of limb, face, and lip movements.

The mirror neuron system is found in both monkeys and humans, and presumably existed in our common ancestor as well. It provides an explanation of why we are so moved and affected by the expressions on another's face, and why the display of body language is such an important part of primate culture (including our own). But more importantly, this could have been the genesis of symbolic language and grammar. It is only a small step from the recognition and generation of emotionally meaningful expressions, sounds, and actions on the one hand, to the abstraction and conventionalization of those expressions, sounds, and actions, and making them stand for something else.

Syntax undoubtedly evolved over thousands of generations and numerous mutations of genes involved in brain growth and differentiation. It almost certainly didn't happen overnight as a result of some 'mega-mutation', or a 'Eureka!' discovery by some long forgotten caveman genius. Theories abound as to how and why syntax actually evolved. One explanation is that it is simply a function of increasing brain size and general intelligence. But that does not make sense because grammar is too modular and specialized to have 'just happened' once we attained sufficient critical brain mass. It is also quite dissociable from general intelligence (there are smart people who can't communicate and good communicators who aren't very smart).

Another group of theories, sometimes dubbed 'environmental theories', suggests that syntax was selected for because it helped us to hunt game, or gather foodstuffs, or make better tools. But there are plenty of animals that hunt very well without having to talk about it. As for tool making, it is possible that there is a causal relationship between linguistic competence and manual dexterity. While tool development did (very slowly) improve during the last two million years in which early hominids presumably developed a universal grammar, there was a sudden explosion of innovative tool making about 50,000 years ago. Corballis suggests that it was the emergence of a vocal apparatus (the lips, larynx, and tongue) physically capable of supporting syntax and symbolic representation independently of the rest of the body that freed the hands to make better tools. But more likely than these general intelligence or environmental theories are the social theories of linguistic evolution. As Steven Pinker writes in **The Language Instinct**:

...it doesn't take that much brain power to master the ins and outs of a rock or to get the better of a berry. But outwitting and second-guessing an organism of approximately equal mental abilities with non-overlapping interests, at best, and malevolent intentions, at worst, makes formidable and ever-escalating demands on cognition. And a cognitive arms race clearly could propel a linguistic one. In all cultures, social interactions are mediated by persuasion and argument. How a choice is framed plays a large role in determining which alternative people choose. Thus there could easily have been selection for any edge in the ability to frame an offer so that it appears to present maximal benefit and minimal cost to the negotiating partner, and in the ability to see through such attempts and to formulate attractive counterproposals. [**Pinker, pg 380**]

Not all social interaction is so harshly Machiavellian. The anthropologist Robin Dunbar has pointed out that brain size in various primates correlates quite nicely with the average size of the social group, which, in turn, directly correlates with the amount of time individuals spend grooming one another. Humans have the largest brains per body weight, have relatively large social groups (averaging about 150 family members, friends, and acquaintances) and spend up to half of their time grooming (socializing). Even with an excellent memory, it becomes increasingly difficult to keep track of 150 people and constantly attend to their ever changing needs and demands without the aid of some sort of language, whether gestural, spoken, or written. Dunbar believes that the demands of social cooperation may have been the selection pressure for the evolution of language and universal grammar. Incidentally, after doing extensive field research on the conversation habits and content of a wide range of 'average' people, Dunbar has found that we spend about **two thirds** of our speaking time gossiping (some of us more than others).

I believe that grammar genes and linguistic brain circuits likely evolved in response to the selective pressure of social intercourse, either of the Machiavellian or the grooming varieties (or both). Language was selected for by the reproductive and/or survival advantage it conferred on those individuals who had it. The advantage is clear: the possession of universal grammar allows one to articulate thoughts, plans, warnings, and dreams to one's peers, relatives, friends, and rivals, allowing more effective and efficient allocation of know-how and resources. It is also a fabulous instrument of courtship; a fitness indicator par excellence that one can use to dazzle or bamboozle a future mate. Charming men use their wit to win wives. Charmed women find witty men irresistible. There is no reason to doubt that our ancestors were any different.

Syntax and the Theory of Mind

Much of what we use language for has to do with comprehending and describing the relationships between different objects and agents: physical, biological, psychological, or purely abstract. Some of these relationships can become very complex. Imagine trying to formulate the phrase, 'this is the cow with the crumpled horn that tossed the dog that worried the cat that killed the rat that ate the malt that lay in the house that Jack built' without words. Words and their arrangement into grammatical phrases allow us to capture and convey the subtleties of complex situations with surprising ease. This is made possible by a device called recursive logic. It is the mental act of taking a nugget of information and embedding it into a larger nugget of information, which, in turn, can be embedded in an even larger nugget of information, and so on. Technically, there is no limit to the extent of recursion we can imagine and produce linguistically other than rules against awkward grammar, the limits of short-term memory, and the attention span of the listener. To quote Pinker, 'short term memory is the primary bottleneck in human information processing.'

Recursive logic is a central component of universal grammar. All languages from Gaelic to Greek, Farsi to Finn, ancient Sanskrit to Modern American Sign use it in one form or another. In fact, the ability to think in terms of recursion may have been the first step in the development of syntax. Recursive logic is also a key component of a related, though somewhat distinct, human ability, theory of mind. TOM is a mental 'trick' which allows the thinker to conceptualize an agent (a person, animal, or thing) as possessing an attitude (a belief, desire, or intention) **independent of physical reality**. For example, 'Ulrike believes that the box is full of candy [even though it is really empty]', or 'Nigel pretended that he was an astronaut [even though I know that he knew that he was not]'. TOM statements such as these utilize the same recursive logic structures we encountered in embedded syntax. The difference is that TOM involves the additional element of **mental states**.

Much research has been done on the syntactic/grammatical competence of autistic individuals. While many of them do rather poorly, it seems that they are at least on par with non-autistics with similar IQ's. Recall that most autistics are mentally retarded or have below average IQ scores. In addition, a substantial subset of autistics, those with 'high functioning autism' (HFA) and Asperger syndrome (AS), have normal grammatical ability. However, all autistic individuals have some degree of limitation on TOM tests. On the other hand, there are individuals (SLI and left frontal stroke patients, for example) who have serious deficits in syntax, with otherwise intact TOM. Syntax and TOM, although related through the use of recursive logic, are dissociable.

It is not clear when TOM may have evolved. But as noted in chapter 3, we have a pretty good idea as to where it may reside. Imaging studies done on experimental subjects performing TOM tests suggest that the critical locus is in the right prefrontal cortex. This roughly corresponds to Broca's area, located in the left prefrontal cortex. Additionally, as I 'pointed' out, both areas contain the mirror neurons implicated in the direct matching hypothesis. It is very possible that both TOM and syntax evolved simultaneously on opposite but corresponding parts of the brain, as our early hominid ancestors learned to capitalize on some serendipitous genetic mutations that enabled recursive, embedded thoughts to occur in their brains.

Pragmatics

Autism is a disorder of social relatedness and communication, but not necessarily one of speech. While most autistics don't communicate in a normal reciprocal fashion, they can talk. The articulation of words (phonemes), the comprehension of their meanings (semantics), and the ability to use them grammatically (syntax) can be generally intact. The profound autistic failure in communication involves another aspect of language: pragmatics.

Pragmatics refers to the social use of language. It is where syntax and semantics meet TOM. The tone and formality one uses to address one's boss, a first date, one's mother in law, or an unruly child are all pragmatic uses of language. Knowing when to ask a personal question, when to take turns in conversation, and realizing when the opposite party wants to change the subject are all contingent on understanding how we use language, not just to convey bare information, but to negotiate complex social contracts with others. In contrast to phonetics and syntax, which deal with the construction of sounds and muscle twitches into words, phrases, and sentences, and semantics, which is concerned with attaching meanings to words, pragmatics is a higher-level process.

We can think of pragmatics as the integration and social application of the three key cognitive 'modules' whose defects lie at the heart of autism: TOM, executive function (EF), and central coherence. In addition, the vast databank of personal experience (stored as long-term memories) is used to find 'pragmatic' solutions to pending social problems. Whenever we have a conversation with someone, it's usually not just talking about the weather (even when we are talking about the weather). Rather, we enter into a sort of unwritten 'contract' with the listener stipulating that we are to provide something relevant and useful for them. We reach into our mental toolbox and deploy the TOM module to find out what it is the listener secretly wants, and use the EF module to formulate a plan for delivering a concise package. We open the long-term memory storage module for the building blocks to construct the message. Finally, we use central coherence to tie the package neat and tight. Stephen Pinker pens it eloquently:

'…The act of communicating relies on a mutual expectation of cooperation between speaker and listener. The speaker, having made a claim on the precious ear of the listener, implicitly guarantees that the information to be conveyed is relevant: that it is not already known, and that it is sufficiently connected to what the listener is thinking that he or she can make inferences to new conclusions with little extra mental effort. Thus listeners tacitly expect speakers to be informative, truthful, relevant, clear, unambiguous, brief, and orderly. These expectations help to winnow out the inappropriate readings of an ambiguous sentence, to piece together fractured utterances, to excuse slips of the tongue, to guess the referents of pronouns and descriptions, and to fill in the missing steps of an argument…

'It is natural that people exploit the expectations necessary for successful conversation as a way of slipping their real intentions into covert layers of meaning. Human communication is not just a transfer of information like two fax machines connected with a wire; it is a series of alternating displays of behavior by sensitive, scheming, second-guessing, social animals. When we put words into people's ears we are impinging on them and revealing our own intentions, honorable or not, just as surely as if we were touching them. Nowhere is this more apparent than in the convoluted departures from plain speaking found in every society that are called politeness. Taken literally, the statement "I was wondering if you would be able to drive me to the airport" is a prolix string of incongruities. Why notify me of the contents of your ruminations? Why are you pondering my competence to drive you to the airport, and under which hypothetical circumstances? Of course the real intent—"Drive me to the airport"—is easily inferred, but because it was never stated, I have an out. Neither of us has to live with the face-threatening consequences of your issuing a command that presupposes you could coerce my compliance. Intentional violations of the unstated norms of conversation are also the trigger for many of the less pedestrian forms of nonliteral language, such as irony, humor, metaphor, sarcasm, putdowns, ripostes, rhetoric, persuasion, and poetry.' [**Pinker, <u>the Language Instinct</u> pg 228, 230**].

Pragmatic failure in autism

I believe pragmatic failure offers a parsimonious explanation of the myriad linguistic idiosyncrasies in autism, from pronoun reversal and echolalia, to stereotypical monologues and difficulty comprehending jokes. I will discuss each of these in turn. At the root of all these anomalies lies an inability to synthesize or analyze intentional messages.

Knowledge does not exist in a 'vacuum' inside people's heads. Once acquired, it becomes personalized to reflect the goals and desires of the possessor. For instance, knowing the opening hours of a store may be remembered only in reference to the time one has chosen to visit it. Much of this 'personal' connection comes from the emotional circuits of the limbic system. It is this 'pragmatic' nature of knowledge that imbues thought and language with its intentional, 'arrow-like' quality. Rather than bits of random informational flotsam and jetsam floating around in a sea of neural circuits, we genuinely believe our thoughts to mean something in the greater scheme of things; they are somehow relevant. And when we speak or write or gesture to others, we intend to convey something of relevance to them. Relevance and intent in thoughts and actions are largely conveyed through linguistic pragmatics, which itself is made up of central coherence, executive function, and TOM. These are exactly the tools that autistic people lack.

Repetition and the lack of spontaneity

We will now take a grand tour of the communicative abnormalities found in the autistic spectrum. I will start with echolalia (the repetitive parrot-like mimicking of heard speech). This phenomenon is not specific to the autistic disorders; it is common in Tourette's syndrome, senile dementia, mental retardation, and other pathological conditions. Even neurotypicals sometimes display it. Have you ever found yourself mindlessly repeating the lines of a crummy song or some catchy phrase, especially when bored, stressed, or preoccupied by something else? Chances are, you have. Echoing of speech is similar to repetitive thoughts and actions in general, as discussed in chapter 5, in terms of EF deficits. Without an effective 'central control' over the generation and frequency of neural activity, the default pathway is simply to continue whatever is going on. Once something is started, it tends to continue moving in that trajectory. Likewise, if nothing is happening, nothing tends to happen. This 'inertia' principle of thought and action is a simplistic analogy, but I believe that it can still be usefully extended to language.

We know that central control is weak in autism. Therefore, we should not be surprised to find repetition of language as well as of thoughts and actions. The units of thought, action, and speech that are repeated can vary from the very small (individual phonemes or sounds) to midsize (complete words and phrases), to very large and intricate (whole passages), depending on the type of autistic personality.

Stress, boredom, or preoccupation are often the triggers for many repetitive behaviors such as mumbling, fidgeting, pacing, smoking, and dancing in both neurotypicals and autistics. This may indicate that repetition is the default pattern when there is no proper suppression from higher centers. These patterns may be generated by subcortical regions, such as the basal ganglia and cerebellum, which are subject to suppression from the prefrontal cortex. When higher centers are otherwise occupied, overwhelmed, or 'down-regulated' in response to stress, fatigue, or boredom, the repetitive patterns emerge. In autism and other disorders marked by repetitive behavior, cortical damage or its disconnection from the subcortical areas may lead to loss of suppression of the lower pattern generating regions.

The flip side of echolalia and repetitive behavior is communicative apathy and indecision. These, too, are common in autism. The pattern generators of autistic people depend strongly on the environment, and are unable to pattern thoughts and behaviors when the environment provides little information, or only inconsistent, unpredictable input. This explains the autistic's lack of spontaneity and creativity as well as the general appearance of 'apathy'. But I don't think apathy is the correct term for what autistics really feel inside. Personally, I often find myself at a loss for what to say or do, usually in unfamiliar social situations. I generally say nothing at all. But it is not that I'm bored or that I don't care. I really do want to say something; I really do want to be social. It is as if the constant barrage of unfamiliar, unpredictable and ever-changing sensory input 'jams' the system. This, in turn, produces increasing anxiety, which can start another cycle of echolalia and repetition.

Uta Frith believes that echolalia may be a manifestation of a loss of central coherence. She writes:

'...echolalia appears to be a glaring manifestation of detachment between more peripheral processing systems and a central system that is concerned with meaning. The autistic child selectively attends to speech and translates heard speech proficiently into spoken speech. However, this processing seems to bypass involvement of central thought. Echolalia demonstrates how end-products of sophisticated information processing can go to waste by not being interpreted by yet higher-level processes. Though they are perfect phonological, prosodic and syntactic units, these products do not become part of global meaning. Instead of becoming tributaries of a mighty river they are streams running into sand.' **[Frith, U. <u>Autism: Explaining the Enigma</u> pg 124]**

This idea is not totally incompatible with my model. But the idea of a 'central interpreter' is somewhat misleading. I think it is more correct to say that the higher levels (cerebral cortex) **modulate and suppress, rather than interpret**, the repetitive default patterns in the lower levels. This is more a function of EF and working memory, than of central coherence.

Pronoun confusion

Another well-known autistic idiosyncrasy is pronoun confusion. It is not uncommon for autistics to say something like, "you want candy" when they mean that they want candy. Pronoun reversal is not a consequence of limited intelligence or memory, nor is it secondary to difficulties in distinguishing individuals. These same autistics will often correctly use proper names to describe characters in similar situations. Pronouns are a special case of symbolic referent because their use is **relative** to the perspective of the speaker and listener. The speaker's 'I' in "I want candy' becomes the listener's 'you' and vice versa. This perspective dependent role taking becomes difficult to handle without abstraction ability and recursive logic. These mental tools allow one to tag different meanings on the same thing, depending on different points of view. Pronoun confusion stems from some of the same problems that also underlie defects in TOM and syntax construction.

Prosody

TOM does play a significant role in the higher pragmatic functions of language. One of these is prosody, or the 'tone of voice'. Prosody encompasses volume, pitch, speed, and enunciation of spoken words. Most of us use some degree of prosody, whether we know it on not, depending on the circumstances and the listening audience. We almost instinctively know when and how to speak deferentially, aggressively, imploringly, tenderly, or sarcastically, and also know when we are being spoken to in these ways. Some people in certain cultures, like Italians, are more emphatic and dramatic than others in their use of prosody. But autistics, regardless of sex, race, or cultural background, are almost universally 'prosodically retarded'.

Contrary to popular belief, not all autistics talk in a robotic monotone (though some do). Most of them are capable of quite a range of tones from slow and articulate, to rushed and excited, to loud and aggressive. The tone of voice is largely driven by emotional input; when we are sad or happy or frightened, the corresponding limbic activation modulates our thoughts, acts, and language. Whenever we feel sad, we sound sad; when we feel excited, we sound excited. We create language to match our emotions. Autistics are no different. They experience and articulate all of these emotions in terms of prosody. The problem is that the prosody is usually inappropriate to the social situation. Many observers have noted that autistics often take the same matter of fact approach whether talking to a stranger about sexual intercourse, or talking to a nine-year-old about quantum physics. They may know what they're talking about, and how to articulate the words and sentences, but they don't seem to know how or when it is appropriate to say it.

How we say what we mean is not just a matter of stringing thoughts into representative sound, sounds into words, and words into grammatical phrases. It also involves knowing why we want to say what we mean to say, what makes us think the listener may want to listen to us, and how to tailor what we want to say so as to have the maximum intended effect. This requires insight into our own and other minds. Being pragmatic speakers, listeners, writers, and readers requires us to be proficient mindreaders as well.

In autism, the neural circuits that subserve the interpretation of others' mental states (which are invisible) based upon their eye movements, body language, and intonation (which are visible or audible) are defective. When the TOM module is broken, there is no shortcut for figuring out other minds. Autistics are, in Simon Baron-Cohen's terms, 'mindblind'. They can still glean an idea of what others may be thinking based on more concrete evidence (what other people actually do or say), but this is a much less effective way of interpreting mental states. It is perhaps for this reason that autistics so often display 'flat', uneven, or generally inappropriate prosody quite independent of their overall intelligence.

To me, the really surprising thing is not that autistics don't speak with proper prosody, but that neurotypicals actually do. Imagine teaching a computer to 'speak' like a human. Advanced speech synthesizers can do a pretty good job of putting sounds together into recognizable words. Computerized dictionaries can be programmed to store prodigious numbers of terms and definitions in multiple languages. Computerized grammar checkers (like the one I'm using right now) do a fair job of analyzing syntax. Voice recognition and dictation software can parse out human speech if we speak slowly enough. But no computer yet, no matter how powerful, has been made to successfully generate human speech with the prosody we routinely and casually give it. Not surprisingly, the machines actually sound like machines, or Mr. Spock from 'Star Trek': logical, but bereft of emotion. Autistics can go a little further and match their internal emotional state with their thoughts and language. But they fail to bridge the next crucial gap between their own emotions and desires, and those of their listener.

Monologues

HFA/AS individuals are sometimes very articulate and verbose, at least in familiar and comfortable social settings. For example, I find it very soothing and invigorating to discuss military history with my girlfriend, or geography with children. I enjoy doing so, not necessarily because I want to teach them something (though I am flattered when they learn), but rather because I like to 'hear' myself 'think'. Going off on a tangential monologue about the Ottoman/Venetian struggles for control of the Eastern Mediterranean in the Sixteenth Century is like taking a hot bath at the end of a long day or getting someone to scratch the part of my back I can't reach. It just feels really good to know someone is listening unconditionally and uncritically!

But this is not really communication. It is a one-way verbal exercise. The standard rules of conversational etiquette are flouted. There is no conversational turn taking, no circumspection of topic, no regard for the listener's receptivity or even interest in what one is talking about. The autistic monologue is pure stream-of-consciousness; it is honest and unconstrained by social boundaries.

I think the autistic monologue stems from a combination of TOM and EF deficits. First, TOM enables the listener to figure out what the speaker means to say, and the speaker to figure out what it is the listener would like to hear. 'Honey, I have a headache tonight' sometimes means more than the sum total meaning of its words. People use TOM to understand intentions, and to generate behavior designed to effect a desired response in turn. Without TOM, all speech becomes literal. Getting people to do what you want and doing what others want is only possible by the most direct brute-force commands. So autistic monologues don't respect the listener's interests. As a result, people often find HFA/AS individuals rude or insensitive (as well as socially awkward) and avoid conversing with them if they can help it. Unfortunately, these misunderstandings only work to further isolate the autistic individual. It is really a vicious cycle.

Secondly, EF problems are perhaps even more important in the genesis of the autistic monologue. The inability to shift from one tone of voice to another (prosody) depending on the type of audience leads to a pedantic, or overly matter of fact style of speech. Children with Asperger syndrome may appear very cute and charming, hence the term 'little professors', but in adulthood, their speech sounds very awkward and even condescending. Failure to change the train of thought leads to persistent repetition of the same story regardless of personal or social context and priority.

Finally, limitations in working memory often prevent these monologues from sounding 'coherent'. Instead of a narrative with an introduction, body, and conclusion or main point, autistic stories are more like concatenations of related facts and observations tangentially leading away from the center, perhaps only to later cut back and repeat the process anew.

Fiction

It has been pointed out that HFA/AS people often like to read books, sometimes obsessively so, but that they much prefer nonfiction. This is certainly true in my case. I personally own about a thousand books, of which perhaps twenty percent are works of fiction. Of those, I may have read less than half. The fiction I have read was either assigned in school, or works of science fiction that tend to be heavier on the 'science' than on the 'fiction'. In contrast, my girlfriend, who is a molecular biologist, reads almost entirely fiction in her spare time.

Why do autistics hate fiction? One guess is that it involves characters whose actions and thoughts are only partly explained by the author. The reader has to fill in the rest by reading 'between the lines' using the TOMM. Dostoevesky never comes out and simply tells us exactly why Raskolnikov murders the old woman; we have to figure that one out on our own. If he had told us, **Crime and Punishment** would have been a much shorter novel, and much less of a great classic. Autistics often find themselves reading fiction 'in the dark', and fail to share in the joys of discovery with their neurotypical bibliophilic counterparts.

A second reason for hating literature is that the number of characters and plot complications can get quite out of hand, given the autistic limitations on working memory. I generally do better with short stories involving a few characters and a simple plot, than with Dickensian or Mitchneresque tomes. It's simply not fun to read when you're trying to keep the cast of characters and events straight all the time. It certainly doesn't help when some of the events turn out to be figments of some character's imagination (as well as the author's) and when things are told in flashback. It's hard to believe that people actually enjoy reading books like this.

A third reason is that fiction, along with biographies and, to some extent, histories, deals largely with human drama. Their contents describe men and women whose actions are usually driven or thwarted by human motives such as lust, greed, avarice, pride, gluttony, jealousy, and other good things. Autistics have a difficult time understanding human motivation in real life, let alone in fictional characters whose actions and lives are often greatly exaggerated for effect. I have a particular aversion to romance novels and psychological thrillers for this reason. On the other hand, I don't mind reading good biographies and histories, perhaps because these human dramas are anchored in particular factual places and times. These things really happened, and cannot be changed. I take comfort in that. Also, most works of history and biography deal with larger social patterns in, for example, early 20th century European colonial geopolitics or English punk rock, rather than dwelling on the inscrutable motivations and ruminations of some fictional chap.

If someone were to give me bundles of money to make a movie, I think I would like to make a great historical war epic, such as Alexander's conquest of the Persian Empire, or the Battle for Berlin. The events depicted would be historically accurate, the action spectacular, the cinematography breathtaking, and the sense of pathos sublime. In all aspects, it would be a great movie, except for one thing. There would be little dialogue. Characters would grunt and scream and shout, and some would even talk (usually in an impeccable foreign tongue like classical Greek or German, rather than have some stupid American speaking English with an ersatz German accent). But they would be talking at each other from a 'distance', rather than talking to the audience. The dialogue would convey no information other than the fact that these characters are talking about what's going on. What you hear is what you get, no more. Talk is really just like another piece of scenery to make the picture more interesting and historically authentic. The beauty of it all comes from what happens in real life, not in someone's head as expressed through language. We wouldn't really know what they were talking about in the heat of the battle, nor would we care.

Some of the best movies made in this 'intentionally detached' style are **2001: A Space Odyssey** and **Barry Lyndon**, both directed by the late, great Stanley Kubrick, who I suspect may have been autistic. In both films, the main ideas are separate from, and bigger than, the minds and designs of the individual characters involved. The dialogue does not drive the stories, but rather, goes along for the ride.

Likewise, if I could write my own songs, I would concentrate on melody and harmony, but not on the lyrics. This is why I prefer easy listening or foreign language radio stations. I can appreciate all the pretty sounds without the words getting in the way.

The books I really enjoy reading are philosophies and general science. They deal with concepts and objects that obey the laws of physics and biology, largely immune to the caprices of human whimsy. Even when I can't comprehend them, I have always held the works of great scientists, mathematicians, and philosophers in higher esteem than those of the great writers and dramatists. For me, there is a comforting consistency in a Newton or Einstein that is missing from a Shakespeare or Joyce.

When I do read fiction, I prefer works that are simply collections of great ideas cloaked in the form of interesting dialogue, rather than collections of great dialogue which just happen to involve interesting ideas. My favourite novelist is Kurt Vonnegut. His brilliant novels are really philosophical and social commentary on the human condition. The characters and dialogue are funny and enjoyable in themselves, but they really stand for abstract concepts.

I am in the minority camp. Most men and women of every culture throughout history (and long before it) have reveled and rejoiced in creating and appreciating works of fictional discourse for their own sake. From mythic sagas told around hypnotic flames of smoky campfires, to Classical Greek dramas staged in open air amphitheaters on the Attic Plain, to the modern spectacles of Broadway and Hollywood, people want to be told lies about other people and places that never existed, when so much of the real universe lies undiscovered all around them. Why?

It has been said that man has an unquenchable thirst for knowledge that has compelled him to create the edifice of modern science and technology. But I believe that he has an even more unquenchable thirst for belief that has compelled him to create endless and sometimes sublime works of fiction and collective delusion. This urge to ascribe intention, agency, and consciousness to dead people, animals, inanimate objects, natural phenomena, and figments of the imagination is probably innate. Three-year-old children will treat stuffed animals and abstract geometric figures on television monitors as living characters, without any adult encouragement. All of this was made possible by a TOM module that evolved by being useful for our ancestors' survival, but was later co-opted for quite different purposes. Our autistic ancestors lacked this module, leaving their descendants without an appreciation of movies, song, literature, and poetry.

Belief

This brings us to the question of the 'types of belief' that one can possess. It is reasonable to suspect that in our ancestral environment, there was an advantage to having precise judgment (of distance, speed, acceleration, weight, time, temperature, etc.). But there were others for whom the world was not all black and white. Truth depended on context and perspective as much as on the contents of sensory experience or logical thought. For them, there were **shades of belief**, rather than simple truth and fiction. This dichotomy exists today. Some people are born skeptics, others are hopelessly gullible. Some believe in magic, sorcery, and religion, others are atheist or scientific.

Autistic people tend to have a narrower window of belief. They believe in what they see and hear and feel, and also often what they are told, whether true or not. But whatever they come to believe becomes rigid doctrine. Most people are more flexible in their willingness to believe. Some things are held very 'true', such as the unconditional love for a child. Other beliefs may be strong, but somewhat malleable in certain situations, such as loyalty to one's nation, religion, or ethnic group. Still other beliefs are very tenuous, such as the temporary and usually voluntary suspension of disbelief whenever we read a novel or step into a cinema. We may lose or gain belief in certain things as we get older (but not necessarily wiser), such as the idea of a tooth fairy or a protective god. Autistic belief is much more 'all or nothing'. Something is either true or not. The concept that belief can come in shades or be momentarily suspended is alien to them. That is not to say that autistics only believe what is true or that all autistic beliefs are true. 'Magical belief' is quite common in autism. Autistics are also notoriously gullible and often victimized by scam artists. But the point is that they hold rigidly to whatever they believe to be true. This, perhaps, explains their dogmatic, self-righteous attitude.

Laughter

We come to perhaps the oddest feature of autistic language: the literal interpretation of figurative speech, including metaphor, satire, sarcasm, irony, and especially, humor. But when I think about it, the really odd thing is the nature of humor itself, and how neurotypicals are so good at decoding it. Jokes are judged by the creativity of their premise and the surprise of their punchlines. The wittier and more intricate the buildup, and the more unexpected or shocking the punchline, the better the joke is said to be. All good jokes in every language seem to share these basic qualities.

Why are we amused by good humor? Why do we find irony ironic? Why do we say something is 'bitingly sarcastic'? What purpose can smiling and laughter serve? These are some really tough questions confronting the evolutionary psychologists. Let's analyze them in finer detail. Irony, satire, sarcasm, and jokes all share a common basic internal 'structure': a logical sequence of syntax and meaning comes into abrupt contact with another totally different and often-inconsistent train of thought (the punchline). The only way to make sense of the anomaly is to adopt a different, unintended perspective, which will allow a consistent interpretation of both sets of events. The humor comes from 'shifting to' the new perspective and 'getting' the joke. The more extreme or novel the shift, the more clever the humor.

The evolutionary psychologists have come up with an ingenious explanation for why this is so. It goes something like this. The brain has evolved to expect certain contingencies. Things in motion tend to slow down; things in the air tend to fall to the ground; old people tend to get sick and die; fierce animals tend to attack when hungry, and so on. When expected things happen, we usually don't notice them simply because it's expensive and wasteful of mental and biological energy to pay too much attention to them. **Unexpected things** are what we need to be vigilant about, because these are the things that can be very helpful or very harmful. Thus, our sensory systems and brain circuits are designed to pick up anything anomalous: a bobbing yellow dot in a background of blue, a note out of key, a special face in the crowd, an off-color comment. When it does, the limbic system, especially the amygdala, is activated to set up the 'fight or flight' response. However, it would be counterproductive, and sometimes downright harmful, to have one's hair stick up on end and break out in a cold sweat every time someone notices a typo, or hears a slip of the tongue. For these situations, we need a 'false-alarm detector', a signal that gives us the 'all clear'. There may indeed be a neural circuit that aborts the autonomic (and automatic) sympathetic nervous system response when we realize there is nothing to fear but fear itself. This false-alarm detector is designed to pre-empt expensive and possibly self-injurious activity by replacing the fight or flight response with an outwardly visible smile or a laugh, and the internal sensations of happiness, amusement, and general well being. Both responses are products of the subcortical limbic system, the same system that makes both anger and laughter largely automatic and difficult to suppress (or fake) without effort.

The neuroscientist V.S. Ramachandran had proposed that the smile, as a 'weaker' form of laughter, may have evolved from an 'aborted' grimace originally designed to scare away strangers by the threatening display of one's canines. When one realizes that the stranger is actually a friend, or that he really is a stranger, but that it would be inappropriate to grimace at him, one simply smiles (whether genuine or not).

Similarly, laughter, whether elicited by a good joke or a good tickle, is a sub-conscious limbic display of an aborted defense response. Stephen Pinker and other evolutionary biologists believe that this false alarm has been co-opted for deployment in social situations where there really is a potential threat. For example, in a boss-employee, or ruler-subject relationship, where open hostility would incur prohibitive costs for all concerned. This explains why so many jokes are 'put-downs' designed to mock or denigrate the authority figure short of actual assault.

Humor and laughter can also be directed inwards. By learning to take our affairs and ourselves lightly and occasionally laughing at our mistakes and misfortunes, we spare ourselves additional grief and risk. In a sense, humor is a self-defense mechanism. Although humor has been elaborated and cloaked in cultural conventions (different cultures have different ideas about what is 'funny'), the biological basis for the sensation we get when we hear a good joke is a human universal.

Finally, if humor is an 'anomaly detector' that is co-opted to deal with social and self-generated threats, perhaps it can also be co-opted for other things, including creativity. Ramachandran writes:

'The ability to reinterpret events in the light of new information may have been refined through the generations to help people playfully juxtapose larger ideas or concepts—that is, to be creative. This capacity for seeing familiar ideas from novel vantage points (an essential element of humor) could be an antidote to conservative thinking and a catalyst to creativity. Laughter and humor may be a dress rehearsal for creativity, and if so, perhaps jokes, puns and other forms of humor should be introduced very early into our elementary schools as part of the formal curriculum.' [**Ramachandran, <u>Phantoms in the Brain</u>, pg 206**]

◆   ◆   ◆

The autistic failure to comprehend complex humor (as opposed to slapstick and other forms of physical comedy, which they generally can appreciate) and the subtleties of wit and sarcasm come from a combination of poor EF and TOM. EF and working memory are essential to effective processing of complex humor. Where in the brain does this happen? As I pointed out in chapter 1, there is evidence that 'high-level' integration and retrieval of complex logical thoughts occurs preferentially in the right cerebral hemisphere. If the prefrontal cortex in general is the integration site for emotions, calculation, and coarse decision making, as I suggested in chapter 5, then perhaps the right prefrontal cortex (rPFC) is the grand integration site for subtle and fine informational processing, including jokes and complex humor. The rPFC may also coincide with the locus for TOM, as supported by fMRI studies of autistic/AS subjects.

The autistic failure to appreciate humor does not mean that they don't know what it's like for something to be funny. In the contrary, AS/autistics can often be easily amused. But their sense of humor is quite literal. Because autistics cannot appreciate the metaphorical aspects of humor, they cannot easily use humor to defuse social threats or internal pain. As a consequence, they take themselves and others too seriously and too literally. They attack real and perceived threats head-on, often with painful, self-destructive effects that increase secondary anxiety. And because autistics cannot easily co-opt humor to serve a creative purpose, they tend to lack imagination, often living their mental lives in repetitive, uninspired ruts.

◆   ◆   ◆

'Hey, Ching Chong! You with the chinky eyes, get your four-eyed ass over here! We wanna play with your glasses! Hong Kong Phooey, number one super guy; Hong Kong Phooey, faster than the human eye. Chicka-chong, chicka-chong, chicka-chong-chong-chong…[laughter].' The childhood sing-song still stings. Three or four bullies, some white, some black, occasionally even girls, would surround me and pretend they wanted to see my glasses. I was easy to pick out. There were few other Asian kids in any of the six public and two private schools I attended, and I was the only one to start wearing glasses at the age of six.

Each year, my parents would take me to the eye doctor for the annual check up. Each year, I would get a new prescription, for thicker glasses. It was bad enough that I had such bad vision and thick lenses. It was worse that my parents insisted on buying me the biggest frames to go with them. By the sixth grade, I had the dubious distinction of having the biggest glasses in school.

'All right! We caught you. You can't run away, ching-chong, 'cus your glasses are too big! Now give us those glasses, let's see your glasses!' They would pretend to go blind, squinting like some ridiculous buck-toothed fu-manchu, bowing and saying, 'assoo, very honorable person!' They would try on the glasses: 'Wow! Check out these coke bottles, man! No wonder you chinks can't see good. Do all Orientals have bad eyes?' I begged and pleaded at my blurry tormentors for my glasses back, but they just laughed and threw them around.

◆   ◆   ◆

'In a democratic society, laws are created for the benefit of its citizens. We, as the citizens of our democratic society, are obliged to actively participate in the making of our laws…as citizens we are also obligated to protest against those laws which we find to be unjust…'

The New England sky was a brilliant cobalt blue on a blustery February morning. The wind blew stray snowflakes onto my face as I distractedly made my way past Harvard undergraduates towards Emerson Hall. It's funny to think so now, but at the time they all seemed so mature, almost intimidating, with their crimson scarves and tweed jackets, Sartre and Kierkegaard casually tucked under their arms. I remember wishing that I too could be as culturally sophisticated as they as I rehearsed my embarrassingly pretentious lines over and over:

'In a democracy, the voice of the government is the voice of the people. Because the voices of all people are different, there must likewise be different and distinctive facets to all issues. The prevailing facet will be generated by the voice of the majority, but this majority must not suppress the opinions of the minorities, for this would defeat the very values and standards upon which democracy was first fashioned…'

As a tenth grader angling for a choice college spot in the winter of 1984, I needed a couple of extracurricular credentials to impress the admission officers at Princeton. My options were somewhat limited. My parents were alumni of no American university, so I had no legacy advantages. I was beyond terrible at all sports, so athletic scholarships were out of the question. I would graduate from a competitive private school with forty students per class, so how many of us could possibly get into Princeton in one year? I suppose if I were black or Hispanic, I would have gotten into Harvard or Yale, but the competition among Asians was fierce. Some of us complained about the reverse discrimination of affirmative action, but the fact of the matter was, if you let every Joe Kim or Sue Lee with a 1500 SAT in, half the freshman class of the Ivy League would be Asian (and who wants that?). Our future class valedictorian, Preeti Bharara (who happened to be Asian Indian) was making quite a name for himself in the speech and debate club, winning just about every trophy below the national level. I figured why not give it a try?

I never was much of a public speaker and tended to stay silent in mixed company. But if I felt comfortable with the audience and topic, I could simply 'explode', enthusiastically expostulating until someone mercifully shut me up. And so it was with my experience in high school speech and debate. I stayed clear of categories like extemporaneous and impromptu speech, which involve spontaneous seat-of-the-pants improvisation. My brain was neither fast nor flexible enough to mount a good verbal assault or counterattack, especially in situations that normally provoke anxiety, such as public speech. But if you gave me a topic, or better yet, let me select a topic, do the research, write the speech, memorize and deliver it to an audience, I wasn't too bad. It all came together at the 1984 annual Harvard University sponsored high school speech tournament, where my 'Civil Disobedience' speech won honorable mention in the original oratory division. A few months later, it took second place at the Long Branch High School speech tournament in New Jersey. Preeti took third place.

Delivering a prepared and over-rehearsed speech is one thing, but being a clever conversationalist is something else entirely. I find it relatively easy to talk to strangers: 'Hello, can I help you, sir?' 'Excuse me, I was wondering if you could direct me to the upholstery section.' 'Please undress, put on this gown, open end toward the back, and I will be back to examine you.' The situations are circumscribed, the delivery stylized, and the options, limited. Cocktail conversations and happy hour banter (as well as public debate) are much more problematic. Unless I have a preset agenda or know the listeners very well, I tend to stay mute. Staff meeting situations are in between. While it is public speech, the agenda is less open-ended than in cocktail parties. I usually have a pretty good idea about what's on the table and who's sitting around it. If I can overcome the not so simple matter of anxiety, I pull through.

Despite what some may initially suspect, I am sure my difficulties with communication have nothing to do with my cultural background or education. While I did not learn to speak English until age five, I was fluent within six months of arriving in Brooklyn. I was reading at an appropriate age level within two years, and generally excelling in all subjects, including English, by the fifth grade. This pattern of rapid linguistic adaptation is really nothing special among young immigrant children. I now speak nearly flawless English with what one of my patients recently described as a 'flat Midwestern accent', whatever that means. Incidentally, I have never been to the Midwest.

My parents have now lived in the United States for thirty years but continue to struggle with English. They can understand standard written and spoken English, but their accents, grammar, and word choice are quite imperfect, and will always remain so. This is not a question of education or intelligence, but of natural age related decrease in neural plasticity. I remember as a child, helping my mother draft business letters. I usually insert appropriate English words into their casual conversation. And, of course, they suffer from the butt of so many Asian jokes, the perpetual confusion of the 'R' and 'L' sounds; 'UCLA' becomes 'UCRA' and I am still called 'HEN-NI'. But we know what they mean. Despite the limitations of their syntax, semantics, and phonetics, my parents usually have a clear idea of what it is they want to express. They get the message across whether in Korean (with other Koreans), imperfect English (with Americans), or a hybrid 'Konglish' (with me and my brother). The common sense and social intent of speech (pragmatics) remain very much intact.

On the other hand, there are times when my father's speech becomes strangely derailed. This is quite independent of his accent, grammar, and choice of words, but rather seems to involve a higher 'expressive' level of language. These situations usually happen when he becomes absorbed in telling a story or 'teaching' us something (math, business, the philosophy of life, etc). He lapses into an almost automatic monologue where the thoughts, words and sentences run on and on like a skipping CD. Sometimes you can't get a word in 'edgewise'. At times like this, it seems that my father is not talking with you, but talking at you, and God help you if you interrupt him.

My mother never talks like this. On the other hand, she is somewhat inexact in her speech (perhaps reflecting her thoughts). Even in Korean, she confuses numbers, sometimes by orders of magnitude ('millions' when she means 'thousands' for example). She prefers to use relative rather than exact terms when giving directions or describing events ('the train station is over there, after a couple of turns, you know where it is', or 'just mix a couple of cups of soy sauce and a few spoonfuls of sugar until it tastes all right'). She may have inherited this propensity for inexactness from her mother, or perhaps it is an 'Asian cultural thing'. But my mother's speech is still generally effective because it conveys the 'gist' of the intended message. What it lacks in rigor and specificity, it makes up for in emotive and pragmatic connectivity.

◆   ◆   ◆

My speech is a combination of parental characteristics. Like my father, I have been known to lapse into stream-of-consciousness monologues, catatonically oblivious to the listener's receptivity. Like my mother, I am often loose or vague in my narrative detail, but usually able to convey the gist of the message.

This is unsurprising if we assume that speech and language, the thought contents underlying them, and the neural activity underlying those thought contents, are all to some extent genetically programmed into our developing baby brains. To borrow the title of Stephen Pinker's book, language is an instinct that all humans share through the evolution and inheritance of certain genes. These and other genes give rise to neural architecture and activity, thoughts, and language that characterize what we call 'temperament' and 'personality'. Likewise, from certain defective genes emerges the autistic temperament.

Part II: Memes

We are the products of our genes. Biochemical instructions in DNA sequences contain the information to build the proteins, cells, and tissues that make up our toenails, spleens, and brains. Genes also specify the particular combinations of cells and tissues, skin pigments, hair texture, and personality types that make us unique individuals of a particular species. Genomes really are the blueprints of individuality.

But that's not all. Locked up in our genes, there is a wealth of information containing the history of our ancestors. Just as species evolve over time, so do their genetic blueprints. In fact, it is the evolution of the genes that causes the evolution of bodies, brains, and behavior. We can see plenty of signs of ancestral phenotypes in contemporary organisms. For example, flowers are specialized modifications of ancestral leaves, the three tiny bones of the mammalian middle ear (called the hammer, anvil, and stapes in humans) are specialized modifications of the reptilian jaw-hinge, and the lungs of all land animals are modified ancestral gills. The bat's wings are derived from ancestral grasping hands, and our modest canine teeth are the remnants of much more impressive ones found in our fossilized ancestors. In the last chapter, I proposed that syntax and the TOMM may be specialized modifications of mirror neurons found in our primate ancestors.

Likewise, there are signs of evolution and adaptation within the genes themselves. Genes (and bodies) don't change abruptly, but rather evolve gradually by building upon existing forms. Over millions of generations, a species of fish may gradually evolve specialized lateral fins useful for skipping about in shallow water, but it cannot sprout legs overnight. Similarly, a new mutation in a gene may slightly alter the peptide sequence of its protein product, but it can't produce a radically different protein overnight. Evolution simply doesn't work this way. As Francis Crick once observed, 'God is a hacker'.

There is a relatively new and exciting subspecialty of biology known as evolutionary genetics. Its proponents attempt to piece together the genealogy of species on the tree of life, rearranging the taxonomy based ones in the old textbooks. By analyzing the differences and similarities in the genes common to different organisms, these scientists hope to discover the ancestral links connecting current species and to trace the particular (and sometimes peculiar) paths taken by evolution. For example, we now know through comparative DNA analysis that whales are relatively close cousins of pigs, both of whom shared a common land ancestor that scavenged ancient beaches about 45 million years ago. Whales are more distant cousins of the reptiles, with whom they shared a common marine ancestor that swam in ancient seas some 400 million years ago, and so on. Yet all these creatures have many genes in common, such as the HOX genes (mentioned in chapter 7) that control body segmentation, and the PAX-6 gene, which controls the development of the eye. The more remote the common ancestor, the greater the divergence in the DNA sequence. As Richard Dawkins writes so poetically in his book, **Unweaving the Rainbow**:

'It is only in a very indirect sense that the genes spell out descriptions of ancestral environments. What they directly describe, after being translated into the parallel language of protein molecules, is instructions for individual embryonic development. It is the gene pool of the species as a whole that becomes carved to fit the environments that its ancestors have encountered...It is in this indirect sense that our DNA is a coded description of the worlds in which our ancestors survived. And isn't it an arresting thought? We are digital archives of the African Pliocene, even of Devonian seas; walking repositories of wisdom out of the old days. You could spend a lifetime reading in this ancient library and die unsated by the wonder of it.'

For complex multi-celled creatures like us, the road from mutation to evolution is an extremely long and tenuous one. Mutations are not uncommon. Among the trillions of cells in your body, each with near identical sequences of billions of DNA base pairs organized into some 30,000 plus genes, spread over 23 pairs of chromosomes, there are undoubtedly quite a few mutants. But mutations occur randomly and accumulate slowly. Most are quickly and efficiently detected and repaired by special repair enzymes. The vast majority of those mutations that escape detection occur in areas of DNA that lie harmlessly outside of genes, or in nonessential parts of genes, or in genes that aren't very important in the overall scheme of the whole organism. In the extremely rare instance that a crucial part of an essential protein in a particular cell is knocked out by a mutation in that protein's gene, the cell will likely malfunction or die before reproducing and passing on the mistake to its daughters. But your body has trillions of cells. So what if a cell inside the lining of a sweat gland in your left armpit dies? Certain mutations or 'hits' in several specific genes regulating cell growth and differentiation (called 'oncogenes' or 'tumor suppressor genes') in say the breast or colon can cause the cell to multiply out of control. This is cancer. But for the most part, mutations are almost insignificant from an evolutionary point of view. Mutations become significant for evolution only when four conditions are met. First, they must occur in functionally or structurally significant parts of genes. Second, those mutated genes must be in the 'germline' cells (the sperm and eggs), which are the only cells passed down to the next generation. Third, those mutant germ cells must be lucky enough to successfully fuse with a germ cell of the opposite sex and develop into a sexually mature adult. Finally, the mutated versions of those germline genes must somehow affect (positively or negatively) the survival and/or reproductive success of the mutant offspring. Successful mutations are those that drive evolution by finding themselves in embryos that survive gestation, birth, and infancy. The young host of the mutant genes must successfully woo and mate with another and pass on his or her mutation to the next generation. If the mutation manifests in a trait that by sheer chance confers a slight statistical advantage to the offspring in their environment, evolution can occur. Of course, this slightly fitter mutant may be promptly eaten by a lion or run over by a truck, but on average, such mutants will be more likely to survive and reproduce more of their kind. The environment changes; adaptive mutations accumulate; evolution happens.

We can think of genes and genomes as both blueprints and history books. Just as blueprints and history books are periodically revised and rewritten, so too are genomes. Books are rewritten by authors and editors. Genes are rewritten by random mutations and natural selection. Evolution works at the level of the organism and its environment to select those genes whose mutations are adaptive for the organism's survival and reproduction. With time, those selected genes build bodies and brains that fit the environment in which they live and reproduce. As the environment inevitably changes, there is a constant selective feedback process on the genes. When ancient seas dried up, those genes that built good bodies for swimming were not selected as well as those genes that built good bodies for walking. This feedback process, with environmental change driving genetic and phenotypic evolution over the course of millions of years, is what created and continues to sculpt organisms that are adapted to their environment. This is why fish swim, eagles fly, and tigers hunt as well as they do. In turn, those bodies that swim better, fly faster, or hunt more efficiently get to pass on more of those genes which build even better swimmers, faster fliers, and more efficient hunters. Genes (the 'software') drive the evolution of bodies and species (the 'hardware') through mutation; environmental adaptation drives the evolution of genes. This is an elegant spiral Richard Dawkins calls 'software-hardware co-evolution'. But what about the stuff that goes on in our brains? Why are sensations, perceptions, and memories as accurate (usually) as they seem to be?

Virtual Reality

Genetic and phenotypic evolution usually occurs very slowly. It is easy for anatomists to see physical evidence of our distant ancestors stamped into our bodies. Likewise, it is easy for molecular biologists to find genetic evidence of our distant ancestors stamped into our DNA. But not all environmental change is so glacially slow. Hurricanes, floods, earthquakes, and meteorites strike suddenly and unexpectedly. Potential foes may attack, potential prey may hide, and potential mates may drop a subtle hint without warning, all within a matter of milliseconds. No amount of genetic anticipation, no matter how tight the software-hardware co-evolution, can predict the chaos and vicissitudes of daily life.

This is why nervous systems evolved. A collection of cells that communicate with one another and form dynamic patterns of electro-chemical activity that somehow mirror changing events in the external world in real time is a very useful tool for predicting the future. And being able to predict the future is helpful if one wishes to propagate one's genes. In order to successfully navigate future contingencies, one needs a 'memory' of the past (immediate, recent, and remote) as well as a continuously updated 'model' of the present. These are precisely what brains are designed to do. Thanks to hundreds of millions of years of selective pressure on trillions of individual bodies and brains and on the genes making them, brains are rather good at remembering one's personal (as opposed to ancestral) past and modeling an approximate facsimile of the here and now.

The 'images' you and I create in our minds' eye, whether the memory of a childhood trip to a summer cabin by the lake, the appearance of the letters and words on the paper in front of you now, or the fantasy of a night with a beautiful Czech woman are not really real. They are fabrications, or, in the case of current sensory perception, approximations of a certain 'deeper' reality underneath the surface. These approximations and fabrications are, for most of our practical purposes, good enough by which to live and multiply. If they were not, we wouldn't be here. But despite the high fidelity of our nervous systems, the worlds they create are 'virtual realities'. They can simulate but never replace the real world because their respective substrates are different. One is made of earth, wind, water, and fire, and governed by the laws of physics and probability. The other is made of neurons and synapses governed by the laws of biochemistry and physiology. Nonetheless, as Richard Dawkins points out, the virtual reality of our minds shapes the selection of our genes as much as the physical reality of our ancestral worlds does:

'We move through a virtual world of our own brains' making. Our constructed models of rocks and of trees are a part of the environment in which we animals live, no less than the real rocks and trees that they represent. And, intriguingly, our virtual worlds must also be seen as part of the environment in which our genes are naturally selected...In the case of highly social animals like ourselves and our ancestors, our virtual worlds are, at least in part, group constructions. Especially since the invention of language and the rise of artifact and technology, our genes have had to survive in complex and changing worlds for which the most economical description we can find is shared virtual reality. It is a startling thought that, just as genes can be said to survive in deserts or forests, and just as they can be said to survive in the company of other genes in the gene pool, genes can also be said to survive in the virtual, even poetic worlds created by brains.'

The Triumph of the Gene

The old social science orthodoxy, disparagingly dubbed the 'Standard Social Science Model' (SSSM) by the evolutionary psychologists John Tooby and Leda Cosmides, went something like this. The brain, like the rest of the body, may be subject to the laws of biology and genetics, but the mind that emerges from it is largely independent of those constraints. There are no 'innate' mental modules, instinctive behaviors, or genetic inheritance of personality. The mind of the human infant is a blank slate, a tabula rasa upon which culture and learning etch out thoughts and ideas. This sort of paradigm became very fashionable among the postwar liberal intelligentsia largely as a 'moral' reaction to the racist horrors inflicted by the Nazis and other right-wing fanatics. Proponents of the SSSM, including the anthropologist Margaret Mead, the psychologist Erich Fromm, and the paleobiologist Stephen Jay Gould, were hostile to the idea that something as miraculous and malleable as the human mind could be the product of random genetic mutations blindly selected by the environment. If that were true, where does it leave room for morality, ethics, and free will itself? The answer, they argued, was that culture is a driving force largely independent of biology that shapes our thoughts through language and social interaction. 'Bad' minds are the result of 'bad' culture; good minds come from good culture. If you can somehow transform culture, you can shape thoughts from the 'top down'. Some in this camp even believed that bad parenting or traumatic childhood experiences caused mental illnesses, such as depression, anxiety, and autism. According to the SSSM, thoughts happen in the brain, but are not caused by the brain. They somehow materialize out of thin air.

Unfortunately, cultural determinism (the idea that culture shapes thoughts) and cultural relativism (the belief that no set of cultural/moral standards are superior or inferior to any other—usually applied as an attack on 'Western Civilization') have tainted the thought of many influential social scientists in the latter half of the Twentieth Century. As Stephen Pinker writes:

'The moral equation in most discussions of human nature is simple: innate equals right-wing equals bad. Now, many hereditarian movements have been right-wing and bad, such as eugenics, forced sterilization, genocide, discrimination along racial, ethnic, and sexual lines, and the justification of economic and social castes. The Standard Social Science Model, to its credit, has provided some of the grounds that thoughtful social critics have used to undermine these practices.

But the moral equation is wrong as often as it is right. Sometimes left-wing practices are just as bad, and the perpetrators have tried to justify them using the SSSM's denial of human nature. Stalin's purges, the Gulag, Pol Pot's killing fields, and almost fifty years of repression in China—all have been justified by the doctrine that dissenting ideas reflect not the operation of rational minds that have come to different conclusions, but arbitrary cultural products that can be eradicated by reengineering the society, "re-educating" those who were tainted by the old upbringing, and, if necessary, starting afresh with a new generation of slates that are still blank.' [**Pinker, How the Mind Works, pg 47-48**]

'The confusion of scientific psychology with moral and political goals, and the resulting pressure to believe in a structureless mind, have rippled perniciously through the academy and modern intellectual discourse. Many of us have been puzzled by the takeover of humanities departments by the doctrines of postmodernism, poststructuralism, and deconstructionsim, according to which objectivity is impossible, meaning is self-contradictory, and reality is socially constructed. The motives become clearer when we consider typical statements like "Human beings have constructed and used gender—human beings can deconstruct and stop using gender," and "The heterosexual/homosexual binary is not in nature, but is socially constructed, and therefore deconstructable." Reality is denied to categories, knowledge, and the world itself so that reality can be denied to stereotypes of gender, race, and sexual orientation. The doctrine is basically a convoluted way of getting to the conclusion that oppression of women, gays, and minorities is bad. And the dichotomy between "in nature" and "socially constructed" shows a poverty of the imagination, because it omits a third alternative: that some categories are products of a complex mind designed to mesh with what is in nature.' [**Pinker, How the Mind Works pg 57**]

◆   ◆   ◆

Sociobiologists and evolutionary psychologists, including E.O. Wilson, Noam Chomsky, Robert Trivers, William Hamilton, Donald Symons, David Buss, Stephen Pinker, John Tooby, and Leda Cosmides have paved the way towards a coherent genetic explanation of human nature by largely discrediting the well-meaning, but scientifically wrong ideas of the SSSM. They have argued persuasively, and backed up with considerable evidence, that our minds, like our bodies and brains, are the products of natural selection and genetic evolution. By extension, culture, which is after all the product of the collective human mind, must also have an indirect genetic basis. Some of the great triumphs of evolutionary psychology include the discovery of universal grammar, innate modules for folk physics and folk psychology (TOM), a genetic basis for altruism based on kin selection, an explanation of divergent male/female sexual behavior based on differing parental investment, and, as discussed in the previous chapter, an evolutionary basis for humor.

Most educated people now accept the (once controversial) fact that evolution really happens and that it is responsible for the great diversity of life on earth. Yet they seem to have difficulty extending that logic to human behavior and culture. One problem is that evolutionary psychology ignores what is 'morally right'. This may be true, but I argue that it is not, nor should it be, the goal of science to prescribe what should be done, or how life ought to be lived. There should be a strict separation of science, on the one hand, from ethics and philosophy, on the other. Good science ought to be beyond good and evil.

A second (and more reasonable) objection is that it seems totally incredible that wonderfully complex adaptations like the vertebrate eye or the human mind simply 'arose' through some mindless evolutionary process. But the fact is, all complex things in nature evolved through natural selection. Natural selection is the only process that can account for complex biological design. The only other explanation is that God made it happen. Personally, I don't find this a satisfying explanation.

But human behavior and culture are not simply biological adaptations. They genuinely seem to be the products of individual and collective volition, and therefore, should lie outside the bounds of genetic determinism. We consciously and intentionally choose our standards of behavior. The idea that culture evolves under genetic constraints somehow doesn't sound right. Yet in a way this is all a big illusion. We think we 'choose' what we do because our brains are designed to follow certain contingencies based on external input and internal wiring. The input is determined by our interactions with the physical world around us. The wiring is largely determined by genes and early experiences. We 'feel' as if we are free to choose because it is adaptive and beneficial for our bodies' survival if our minds are constructed in this way. Free will is an illusion that our genes impose upon our brains to make better decisions. We will return to the exceptionally thorny issue of free will in more depth in the next chapter.

Ultimately, our brains and their phenomenological contents, both what we colloquially call the conscious mind and the myriad unconscious processes underneath, are constructed by the genes, and evolve by Darwinian selection. And, as we saw earlier, the contents of our minds, those virtual realities of conscious experience, shape the selection of our descendants' genes just as much as our ancestors' physical realities selected our own genes.

Culture and behavior are linked to genes because they form part of the 'environment' in which the genes' vehicles (our physical bodies and brains) are selected. Language may be the critical link here because it allows the (near) faithful transmission of cultural information from one individual to another in a different place and time.

Universal Darwinism

The contents of our minds: all our perceptions, emotions, memories, opinions, and decisions are constantly created, edited, and re-edited in the circuits of our brains over the milliseconds, minutes, days, and years of our lives. We are born with a full complement of genes etched into most of the cells of our bodies, and, with a few exceptions, maintain this complement throughout life. Perceptible genetic evolution is a process that occurs over thousands of generations, which, in humans, is spread out over tens and hundreds of thousands of years. Yet the thoughts, habits, languages, and customs acquired over the course of a single lifetime profoundly affect the coterie of genes selected for the next generation. How?

The ability of animals to learn is very ancient. Fruit flies and sea slugs can quite easily be taught to avoid noxious stimuli and approach attractive ones. All animals from this level on up to humans have at least some innate ability to learn and, in a few cases, imitate others' behavior. The ability to learn and imitate are largely genetically determined functions of neural circuitry and synaptic plasticity (chapter 8); they are natural. But the actual traits that are learned, such as tool making, polite table manners, dating etiquette, and the ability to speak French, are acquired non-genetically. These traits are dependent on where you went to school, and whom you hung out with as a child; they are cultural.

At the turn of the 20th Century, psychologist James Mark Baldwin proposed that the very ability to learn something useful could be subjected to Darwinian selection. If the trait that is learned (whether conceptual, behavioral, linguistic, or an element of culture) is beneficial enough to affect the survival and reproductive rates of those who are able to learn it, the general genetic propensity to learn or imitate will be selected. As a result, over time, their progeny will evolve to learn and acquire useful traits faster and more efficiently. Thus, genes and 'cultures' evolve together, producing more powerful brains capable of more subtle, abstract, and convoluted thoughts, and the application of those thoughts to generate more complex language, culture, and technology.

In this sense, any and all products of our minds (and the 'mind' itself) are really adaptations that help our genes reach the next generation. As the sociobiologist E.O. Wilson put it, "the genes hold culture on a leash. The leash is very long, but inevitably values will be constrained in accordance with their effects on the human gene pool." The evolutionary psychologists have taken this stance as the cornerstone in their remarkable search for the origins of human social behavior. They have found plausible genetic advantages behind (and possibly explaining) such unlikely phenomena as altruism, racism, male aggression, adultery, polygamy, and incest taboos. Even reproductive losers such as homosexuality, schizophrenia, and autism might be at least partially explained using the concept of 'heterozygotic advantage'. This is a concept we encountered in chapter 6 referring to a gene that in double dose (homozygosity) causes disease or reproductive disadvantage, but in single dose in the relatives of the afflicted (heterozygosity) confers an advantage.

The SSSM maintains that our thoughts are largely sculpted from the 'top-down' by cultural patterns that are independent of genes. The evolutionary psychology paradigm, in contrast, proposes that thoughts are the 'bottom-up' products of genes and brains. These thoughts are then the raw materials from which languages and culture are built. (**figure 1**)

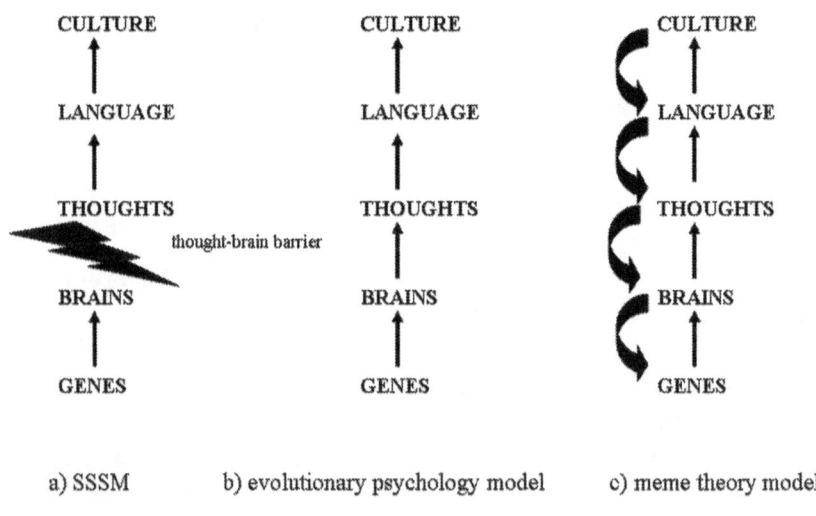

a) SSSM   b) evolutionary psychology model   c) meme theory model

Figure 1

◆   ◆   ◆

Despite all of its remarkable successes, the evolutionary psychology approach has run into some difficulty explaining certain aspects of human behavior. Religion, superstition, music, art, fashions, and fads all seem to serve little or no advantage for the genes. It doesn't matter to your genes whether you happen to be a Moslem, Methodist, or Mormon, prefer listening to Beethoven or Britney Spears, eat fish and raise dogs, or eat dogs and raise fish. While some cultural practices such as circumcision have little effect on genetic fitness, others such as voluntary celibacy, birth control, suicide attacks, and genocidal holocaust are quite harmful to genetic propagation, yet they have survived and sometimes thrived throughout the centuries (although they are markedly more common now). Why? Who benefits? One explanation for things like art, music, fashion, and the habitual over-consumption of intoxicating substances is that they serve as genetic 'fitness indicators' for sexual selection aimed at attracting members of the opposite sex. In his book, **The Mating Mind**, the evolutionary psychologist Geoffrey Miller argues that maladaptive behaviors such as warfare are actually co-opted fitness indicators that have spiraled out of control. He goes so far as to suggest that all art, culture, and even our unique creativity and intelligence, as well as the big brain which makes it all possible, are the aftereffects of women continuously selecting men who can afford to support the most aggressive, expensive, elaborate, extravagant, ostentatious, and decadent habits and lifestyles. Men are therefore forced to go to ridiculous extremes to outcompete one another and impress ever more finicky women. This sexual selection theory of culture is very interesting, and major aspects of it may be true, but it does seem a bit of a stretch of the genetic leash.

◆   ◆   ◆

All complex adaptations, whether biological, neurological, or cultural are the products of Darwinian selection. The fundamental laws of physics, inorganic chemistry, probability, and thermodynamics governing the formation of dense star clusters, great galactic spiral arms, and the chaotic patterns of subatomic particles are profoundly powerful, but too general to create the kinds of complexity found in biochemistry, molecular genetics, neural networks, and human culture. These are emergent phenomena emanating from natural (Darwinian) selection. Genetic selection is one kind of Darwinian selection, but not the only possible kind. Genetic selection and biological evolution depend on the replication of DNA. Other kinds of selection may involve other (very different) replicators. This is the concept of 'Universal Darwinism'. The basic principles of variation, heredity, and selection are the same, but the nature of the replicator is different.

The brain itself may use a kind of 'neural replication' of synaptic 'feedback loop' patterns to generate coherent mental states. The molecular biologist and neuroscientist Gerald Edelman calls this 'neural Darwinism' (see chapter 11). Extraterrestrial life, if it indeed exists, is likely to involve Darwinian selection of some sort with a non-DNA based replication system. If computers can one day be made to self-replicate, perhaps using a 'silicon code', they, too, will be subjected to Darwinian selection. Likewise, cultural evolution might undergo Darwinian selection, independent of the genes, if there is a second replicator.

Memes

At the end of his now classic book **The Selfish Gene**, Richard Dawkins introduced the term, meme, referring to a unit of cultural inheritance. Memes are packets of information: a line of poetry, a bit of political ideology or religious dogma, a recipe for shepherd's pie, the directions on how to copy a gesture, a tune, or a dance step, that are stored in the minds of individual people and then transferred and replicated in others. Memes are selected for replication based on their usefulness or simply on their opportunistic ability. Quite a few animals are able to mimic gestures and behaviors. Some of them, such as the vervet monkey, even use gestures to represent predators. But only humans transmit and reproduce meaningful patterns of behaviors readily and profusely using the media of spoken, written, and now electronic languages. Language allows us to encode the instructions for doing or making something (the meme or 'memotype') into a unit of 'high fidelity, fecundity, and longevity' (the expressed meme 'phenotype'). In this sense, memes are analogous to genes: they are the units of cultural heredity that undergo spontaneous variation, replication, and selection and are subject to competition and cooperation with other memes. Memes are a second replicator.

The influential philosopher Dan Dennett believes that the human ability to propagate memes using language has opened up a whole new landscape of cultural 'design space': an arena for conceptual and behavioral patterns not bound by biological or genetic constraints. He writes as a critique of E.O. Wilson's gene centered view of culture:

'...Wilson's leash is indefinitely long and elastic. Consider the huge space of imaginable cultural entities, practices, values. Is there any point in that Vast space that is utterly unreachable? Not that I can see. The constraints Wilson speaks of can be so co-opted, exploited, and blunted in a recursive cascade of cultural products and meta-products that there may well be traversable paths to every point in that space of imaginable possibilities. I am suggesting, that is, that cultural possibility is less constrained than genetic possibility. We can articulate persuasive biological arguments to the effect that certain imaginable species are unlikely in the extreme—flying horses, unicorn, talking trees, carnivorous cows, spiders the size of whales—but neither Wilson nor anybody else to my knowledge has yet offered parallel grounds for believing that there are similar obstacles to trajectories in imaginable cultural design space. Many of these imaginable points in design space would no doubt be genetic cul-de-sacs, in the sense that any lineage of H. sapiens that ever occupied them would eventually go extinct as a result, but this dire prospect is no barrier to the evolution and adoption of such memes in the swift time of cultural history. To combat Wilson's metaphor with one of my own: the genes provide not a leash but a launching pad, from which you can get almost anywhere, by one devious route or another. It is precisely in order to explain the patterns in cultural evolution that are not strongly constrained by genetic forces that we need the memetic approach.' [**Dennett, "The evolution of culture", the Charles Simonyi Lecture, Oxford University, Feb 17, 1999**]

Dennett makes an interesting analogy between the relationship of memes and their human hosts on the one hand, and parasites and their biological hosts on the other. Some meme expressions, such as the practice of washing hands after using the bathroom or saying 'excuse me' after bumping into a stranger on the subway benefit both the host (genetic advantage) and the meme. These are 'mutualistic memes'. Other meme expressions, such as humming a tune while waiting for the elevator or saying 'gesundheit!' when someone sneezes, don't confer any genetic advantage. These are 'commensal' memes. Finally, some meme expressions are acutally harmful for the host's genes. These include the previous examples of celibacy, drug abuse, and warfare. These are 'parasitic' memes. It is easy to see why evolutionary psychologists would classify mutualistic and even commensal memes as cultural artifacts on genetic leashes. It is much harder for them to explain away parasitic memes using a 'gene's eye view'.

Now it is still true that our physical bodies are created and maintained by genetic instructions. If a wildly successful but extremely pernicious parasitic meme, such as an escalating global thermonuclear conflagration or uncontrolled global warming, managed to infect and destroy the human race, then the meme itself would die out (there would be no more human minds and bodies left to infect). But generally, like all 'good' biological parasites, most parasitic memes are not that virulent or harmful to the host's genes. Also, genes and memes co-evolve. Genes evolve to 'track' the evolution of memes so that their mutual host (the person who passes on the memes and the genes) stays alive and healthy enough to continue replicating the memes (through linguistic discourse) and the genes (through sexual intercourse). The problem is that memes reproduce and evolve much faster (seconds, days, and years) than genes ever can (decades, centuries, and millennia). As a result, genes eventually lose track of the memes. To continue this metaphor, the leash that holds the memes grows longer and longer, until the memes simply break free.

Culture, language, and perhaps consciousness itself are collections of memes that sometimes cooperate and sometimes compete with one another for space in our brains and the opportunity to reproduce by commandeering our communicative organs and taking up our time and energy. Like genes, which cooperate to produce an organism, memes can cooperate to produce a culture.

'Memes, like genes, survive in the presence of certain other memes. A mind can become prepared, by the presence of certain memes, to be receptive to particular other memes. Just as a species gene pool becomes a cooperative cartel of genes, so a group of minds—a 'culture', a 'tradition'—becomes a cooperative cartel of memes, a memeplex, as it has been called. As in the case of genes, it is a mistake to see the whole cartel as a unit being selected as a single entity. The right way to see it is in terms of mutually assisting memes, each providing an environment which favours the others. Whatever may be the limitations of the meme theory, I think this one point, that a culture or a tradition, a religion or a political complexion grows up according to the model of 'the selfish cooperator' is probably at least an important part of the truth.' [**Dawkins, <u>Unweaving the Rainbow</u>, pg 306**]

The History of Memes

The model of cultural evolution promoted by evolutionary psychology is a very useful and often accurate approximation of cultural trajectory up to a point (**figure 2**). The last several thousand years of human history, however, is difficult to model using the rules of genetic advantage. Twentieth century cultural history is almost impossible in this way. To see why, we have to explore how memes diverged from genes.

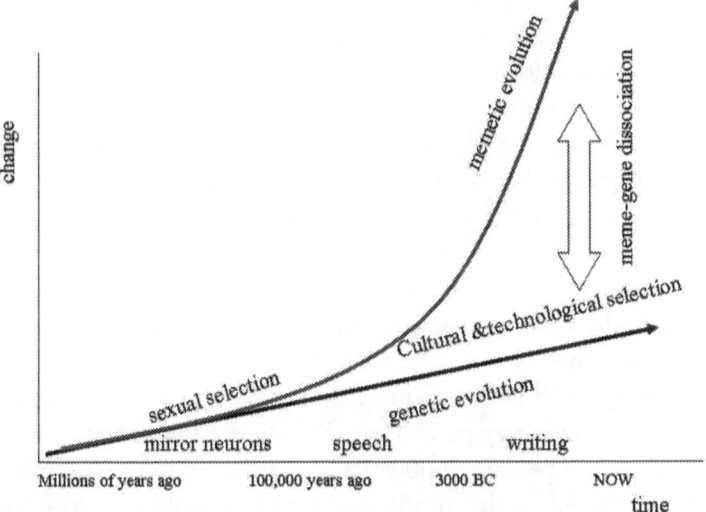

Figure 2: memetic trajectory

The very first 'protomemes' probably started when early hominids learned to imitate facial expressions and hand gestures. Our monkey cousins have the neural machinery (mirror neurons) to do this. We can assume that our common ancestor some 15 or 20 million years ago was doing this as well. This is the biological basis for meme production. But 'true memes' involve the transfer of the instructions for doing something, rather than just the external gesture devoid of symbolic content. What monkeys exchange (protomemes) are like 'junk DNA'; what humans exchange (true memes) are more like real genes. For that to occur, information (software) must be coded into the physical expressions (hardware). This becomes possible only with the emergence of representational thought and recursive logic, the dual basis for Theory of Mind and syntax. The first true memes arrived perhaps two to three million years ago.

At this time, our ancestors were capable of producing true memes, but their general mental capacity ('intelligence') and communicative ability were probably quite limited. As such, the memes that were exchanged were unlikely to successfully propagate unless they somehow assisted biological survival. A better way to make a fire, throw a spear, or maintain a poker face in the midst of danger would be examples. Such memes are of the mutualist type. The genes maintain the memes on a leash; the evolutionary psychological approximation is quite close.

Thanks to some lucky mutation(s), probably including the FOXP2 gene (chapter 7), about 50,000–100,000 years ago, our ancestors presumably developed articulate speech, mastered refined stone tool technology, killed off the remaining early hominids, and left Africa to conquer the world. Spoken language allowed them to transmit ever more complex memes faster and more efficiently without hindering the use of the hands for other things. There was a great co-evolutionary spiral as the mind thought more intricate thoughts, the lips and vocal cords produced more articulate speech, and the hands built more useful structures. This was the beginning of complex human culture. For the first time, memes were unleashed from genetic advantage. Many of them remained mutualist, but many others became parasitic, surviving by banding together with other memes and going their own way far from the protective leash of the genes.

The trajectory of these divergent memes cannot be accurately approximated by the ruler of evolutionary psychology. But during the period from 50,000 or 100,000 years ago to perhaps 3000 years ago (and even today in some fast disappearing cultural islands in the world), the keys to survival and success were largely genetic factors such as general intelligence, physical attractiveness, strength, height, fertility, and so on. These are the things that enable individuals to better feed themselves, woo and win higher quality mates, and have lots of kids just like them. There was not much private property or 'wealth' to accumulate, and therefore, not much for a strong man to bequeath or a weak man to inherit. Status was determined by physical and intellectual prowess, which is largely genetic, rather than by socioeconomic standing, which is largely memetic. That would come later. But for this long earlier epoch, cultural trajectory still closely mirrored genetic evolution.

With the invention of writing, everything changed. The meme-gene divergence started to grow at an accelerated rate. People could collect lots of little memes together on pieces of clay or papyrus or scrolling paper that they couldn't otherwise hold in their heads all at once. Their brains then scanned the mini memes and spliced them together in their minds to create new and more complex 'memeplexes'. Some of them, such as 'the Bill of Rights' or the smallpox vaccine were useful for the genes. Some others like A Midsummer Night's Dream or the 'Gospel according to St John', were not too useful, but enjoyable, or sometimes useful and sometimes harmful. Still others, like the Spanish Inquisition and the breech loading rifle were generally harmful to the genes, but propagated anyway because they were supported by so many other memes (such as the concept of Catholic nationalism or the 'white man's burden') and quite a few genes (the minority of Spanish priests who grew wealthy off confiscated Moslem, Jewish, and Protestant property; the few Europeans who always won the colonial wars; successful weapon manufacturers and merchants).

Memes piled on top of memes, and successful coalitions of cooperating memes—the memeplexes—were passed on horizontally from person to person, and vertically from generation to generation through documents, libraries, and universities. Certain memeplexes, such as agriculture, double entry bookkeeping, stock markets, and mass production allowed some people to amass more than their 'fair share' of resources. Because these people had more than they could personally use in their lifetime, they devised elaborate customs and laws for passing on accumulated wealth to designated heirs (usually their offspring) regardless of genetic fitness. Meanwhile, the great mass of people who failed to amass the wealth, some of whom no doubt had the innate ability to do so if they had been given the chance, were left out. Thus emerged the concept of socio-economic class. Class is largely memetically inherited. Genetic endowments such as natural intelligence, good looks, physical health, and big muscles can sometimes win class status, but often inheritance went to the children of the wealthy. From the rise of civilization in the Fertile Crescent some 5000 years ago at least to the widespread application of meritocracy in a few parts of the Western World, especially Great Britain and the United States starting in the Nineteenth Century, cultural trajectory increasingly mirrored the memes rather than the genes.

Over the past century or so, the pace of memetic (cultural) evolution has been mindboggling. New laws, states, fashions, books, discoveries, and ideas spring up and die in the billions. Storage and transmission has been greatly amplified by advances in communication technology such as the telegraph, radio, photocopier, fax machine, computer modem, cellular telephone, and wireless internet. In fact, biological evolution has in certain ways been slowed down or halted as modern medicine keeps unfit genes alive and effective birth control prevents the spread of otherwise healthy genes. Human genetic evolution itself is now as much under the control of memes as of genes. Natural selection of genes is being supplanted by memetic and genetic engineering. The leash has been reversed.

Ironically, just as the meme-gene gap is expanding as never before, there is a collection of memes that seem to extol the virtues of the genes. This 'retro-memeplex' takes the form of the political philosophy known as 'meritocracy' and, more recently, the scientific movement known as evolutionary psychology. They have become widely accepted in the land that churns out more memes, mutual as well as parasitic, than anywhere else on earth, the United States of America. It is rather odd to think that I live in a country where most people believe that others should be judged not by the size of their father's bank account or by the caste or country club to which they belong (which are largely cultural), but rather by the strength and content of their personal character and their physical and intellectual achievements (which are largely genetic). These ideas, though widely accepted in the wake of globalization of free markets and the uncontrollable spread of American culture, are rather novel in the course of written human history. But we should realize that these too are just memes, and their ultimate acceptance or denial do not currently make much of a dent on the soaring trajectory of the memes.

The New Top-Down Theory

It is now clear that the process and content of our thoughts are shaped as much by culture (via memes and language) as they are by genes (via the physiology and anatomy of the brains they build). The bottom-up paradigm of evolutionary psychology (**figure 1**) should now be amended to include the top-down processes made possible by meme theory (**figure 3**). This is a synthesis of the old culture-based SSSM with the more recent gene-based models. The SSSM neglected the crucial role of genes and innate cognitive modules in the thought process. Meme theory takes them into account as both **bottom-up sources** and **top-down targets** of our thoughts.

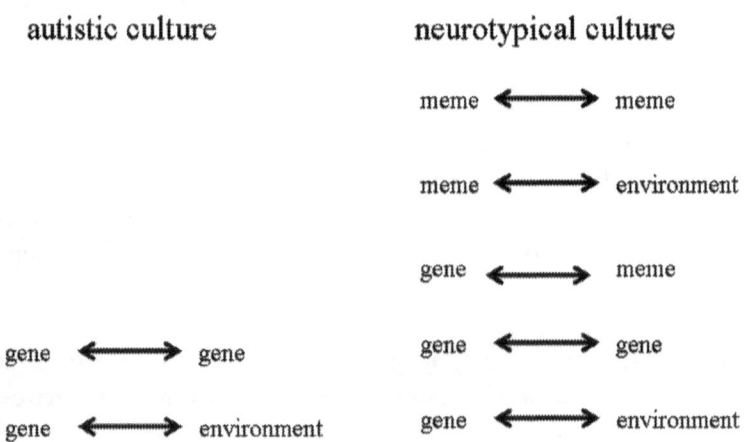

Figure 3: environment-gene-meme interactions

In earlier chapters, we focused on the mind as a product of bottom-up processes and modules such as sensory perception, motor circuits, emotions, memories, executive control modules, TOM modules, and universal grammar. All of these 'emerge' from the dynamic activity of trillions of plastic synaptic connections of billions of neurons, which, in turn, are largely products of genetic code interacting with the environment. This is the 'gene's eye view': bodies, brains, thoughts, language, and even culture are simply (complex) vehicles designed by the genes to make more of themselves through sexual reproduction. The genes sometimes compete and sometimes cooperate with one another (in the form of genomes which make up organisms) in this perpetual and epic process known as 'life on earth'.

To this, we must now add the 'meme's eye view': culture, language, and thoughts are simply vehicles designed by the memes to make more of themselves by social imitation. The memes sometimes compete and sometimes cooperate with one another (in the form of 'memeplexes' which make up human minds) in another perpetual and epic process known as 'human culture'.

Genes working from the bottom up and memes working from the top down often intersect and interact in the middle: the contents of the conscious mind. The evolutionary psychologists have proven that human minds are not blank slates that can be programmed or reprogrammed at will. But it is also quite clear that we are not prisoners of genetic design. Our minds are to some extent malleable and subject to the influence of cultural mores. To quote the cultural anthropologist Richard Nisbett:

'Thus it appears that the assumption that cognitive content is learned and indefinitely malleable and the assumption that cognitive processes are universally the same and biologically fixed may both be quite wrong. Some important content may be universal and part of our biologically given equipment, and some important processes may be highly alterable. The continued existence on the planet of widely different social and intellectual traditions offers an opportunity to learn a great deal more about the fixedness and malleability of both content and process.' [**Nisbett, R., et al, (2001) 'Culture and systems of thought: Holistic versus analytic cognition', <u>Psychological Review</u> 108: 2, 291-310**]

The question is, how far down can memes go?

◆   ◆   ◆

It appears that cultural characteristics can have robust effects down to the level of basic visual perception and attention. Nisbett and his colleagues have done some remarkable cross-cultural studies comparing the way East Asians (Chinese, Korean, and Japanese) and Westerners (North American, Western European, Australian, New Zealand, and South African Caucasians) perceive and attend to stimuli, and make judgments. In one experiment (Masuda & Nisbett, 2001), Japanese and American subjects were shown identical scenes of fish swimming in a pond. When asked to describe the scene, the Americans tended to concentrate on the characteristics of the foreground: the principal fish and what it was doing. The Japanese subjects concentrated more on the background: the other fish, the pond weed, the interaction of the fish with aquatic plants, etc. It seems that the Americans are concerned with getting the 'main point', while the Japanese were more 'gestalt' oriented.

In another series of experiments (I. Choi & Nisbett, 2000), Americans and Korean subjects were asked to comment on narratives with expected and unexpected outcomes. The Koreans were less likely to be surprised by unanticipated outcomes. They were also more willing to accept contradictions in logic or inconstant interpretations than the American subjects.

These studies suggest that East Asians are more 'holistic' in their perceptions and thoughts. Their attention is 'bound' to the circumstances surrounding objects and events rather than to the objects and events themselves. This may explain the stereotypical Asian penchant for finding the 'middle way', compromise, non-confrontation, and fatalistic resignation typified by 'feng shui', loose and flexible interpretation of legal contracts and religious doctrine, and the sometimes overwhelming need to 'save face'. In fact, the Chinese language itself is much more context, rather than syntax-based than the Indo-European languages. Asian children are found to learn verbs faster than they learn nouns, while the opposite holds true for Western children. Even the pictographic nature of Chinese writing reflects the emphasis on relationships and general gist while the alphabetic writing of the West emphasizes precise definitions and modularity.

The origins of cultural differences is a highly controversial arena that goes back to the old 'nature vs nurture' debate. Nisbett and other cultural anthropologists are squarely in the top-down camp. While they may recognize the importance of genes as building blocks of biological and some psychological systems, they maintain that the elements of culture are purely arbitrary human constructs that are passed down and modified as societies evolve over time. One line of speculation is that contemporary East Asian societies reflect the characteristics of the ancient Chinese civilizations from which they arose. Ancient Chinese culture was based on agriculture. The social roles and expectations of the people revolved around the cooperative effort required of farmers. As these people learned to live in tight cooperative social units over generations, their way of thinking, their language, and the beliefs, and even their perceptions of the physical world took on a more 'holistic', collective, 'context-dependent' character. Modern Asian culture has changed drastically in the last century as European Imperialism, communism, urbanization, and now American style capitalism have taken hold; superficially, much of the original 'Asian' character has been lost. But fundamentally, East Asian society is still deeply influenced by cultural patterns set in ancient China.

Western society is largely derived from the Greco-Roman and Judeo-Christian traditions. The two great fountains, the ancient Jews, and especially, the Classical Greeks, were (due to geographic and environmental factors) not primarily agriculture-based people. They were traders and merchants. The social roles and expectations of these peoples depended to a large extent on individual initiative and entrepreneurial resourcefulness. As these early 'Westerners' learned to live in highly competitive social environments over generations, their way of thinking, their language, their beliefs and perception of the physical world took on a more focused, individualistic, and 'object-dependent' character.

There were times, such as the thousand years between the fall of the Western Roman Empire to the barbarians and the start of the Florentine Renaissance, when the characteristically aggressive mercantile spirit of Western Europe was subdued by religious superstition, economic stagnation, and bacterial epidemics. Perhaps at such times, European culture and psychology took on a more 'Asian' character. But fundamentally, and especially in the last few centuries, Western Civilization reflects cultural patterns set by the Classical Greeks.

◆   ◆   ◆

Cultural differences between societies based on ancient China, those based on Classical Greece, or on any other culture throughout history, can be quite large and profound. But, according to the top-down school, they are not genetically fixed. Short of genocide or some other massive sociocultural dislocation (which do happen from time to time), it is exceedingly difficult to get a society to change its character overnight. However, it is quite easy to get an individual from one society to change his or her cultural character (especially if done early in life).

Chinese children born or raised in America act, talk, and dress like other American kids (it does seem that Western children born or raised in Asia don't usually act, talk, or dress like Asians, but this too is likely to have a cultural, rather than genetic explanation). Studies have shown that Asian American children and adults think pretty much like white Americans (or at least show no significant group difference when compared to white Americans on tests of 'figure/ground' perception and social judgment similar to the Nisbett studies mentioned earlier). On the other hand, their parents, who were raised in Asia, usually perform more like native Asians. This cannot be a genetic effect because the second-generation Asian American subjects have Asian parents and thus 'Asian genes', whatever they may be. The cultural change between the generations must be acquired through learning (memes).

Incidentally, however, some people still stereotype all East Asians (whether native, immigrant, or third or fourth generation American and totally assimilated) as passive and effeminate. Like most stereotypes, there is probably at least some truth to this myth. I can think of three possible explanations for this alarming and insulting situation. The first is that the perpetrators are simply ignorant, and learned to characterize Asians from the way they are portrayed in Hollywood movies (which has always tended to portray Asians very unflatteringly and stereotypically) and simply generalize this image to all 'real-life' Asians and Asian Americans. The second is that Native Asians (and immigrants to some degree) actually are somewhat effeminate as a reflection of their culture, but that this wears off as their children and grandchildren become assimilated. A third possibility is that East Asians, whether native or totally assimilated, are, in fact, biologically and psychologically effeminate, regardless of acculturation. The first two explanations, which are meme based and top-down, are more palatable. The third, which is gene based and bottom-up, is much more disturbingly racist. However, we should always strive to keep good science from being swayed by the currents of political correctness, emotional comfort, or whatever happens to be 'morally right' or fashionable at the time. Perhaps more experiments should be done to tease out the relative influence of cultural factors vs genetic factors on unconfirmed 'racial characteristics', such as the myths of East Asian femininity, Caucasian aggression, Jewish intelligence, and black African physical prowess.

Autistic Culture

Nature and nurture, genes and memes, mind and culture, all of these seemingly discrete and definite terms are really just vague approximations of thoughts, collections of symbols on paper, sounds on the lips, linguistic illusions that we have created in our never ending attempt to grasp the essence of what is true and turn it into words, often unsuccessfully. If you really think about it, the mental world in which we live, at the social scale at which we live it, feels smooth and continuous; memories, beliefs, thoughts, urges all melt and blend into one another over time. There is also no fine line dividing genes from proteins from neurons from brain circuits from thoughts from language from culture. It is a continuous ribbon of emergent information. Our minds are constructed from the genes up and from our cultures down. Even the genes themselves are really arbitrary concatenations of nucleic acids whose sequence contains the information and directions for building the next level up. Likewise, memes are random electrochemical activity in the brain whose patterns contain the information and directions for building the next level up. The mind uses language to bridge the gap between memes and culture. Richard Dawkins writes, 'languages evolve to become good at infecting child brains. But the brains of children, those mental caterpillars, also evolve to become good at being infected by language...' Yet we saw earlier that autistic children have limited use of language. They are immune to the viruses of language, and as a result, are impervious to the transmission of memes and culture. In this final section we will examine the memetic limitations of the autistic mind and its implications for the 'culture' of autism.

Just as our minds are products of our cultures acquired through language and memes, cultures are constructions of the human mind working through memes, languages, and human behavior. The degree of cultural sophistication is proportional to the volume and density of memes in the 'memeosphere', which in turn is a function of the amount of brainpower people use to interact and exchange ideas. Autistics have limited social brainpower for several neurological reasons (limited TOM, EF, working memory, and so on), as a result, they are both mentally and culturally handicapped.

People learn the directions for following traffic rules, producing goods and delivering services, bowling, fixing cars, and acting in a rational, civil manner. The directions are memes. They carry out these directions through imitation. This is culture. Memes build culture as they spread through a population of minds through imitation. Culture is the grand arena—the memeosphere—in which the motley panoply of human memes joust about. What the biosphere is to the genes, culture is to the memes.

But I think a complex culture involves more than just the replication of memes. Simple imitation of gestures, sounds, or even entire passages from the King James Bible or the Sports Almanac do not make a culture. The individual must be aware of the underlying significance of the transferred meme; he must know that whatever is imitated means something and represents a component of some larger intention. Reciting pi to 10,000 decimal places or all the nations of the world in alphabetical order is simply imitation for the sake of imitation. Likewise, echolalia, perseveration, stereotyped interests, and monologues are all examples of memetic imitation, and some autistics are indeed very good at it. But these memes are not the ones that build complex culture.

This brings us to the core question of what kind of culture autistics have. This is really impossible to answer for sure, as autistics live within a larger non-autistic ('neurotypical') community—a community whose culture is shaped by the dissemination of robust memes. But autistics don't make and spread these 'strong' memes. The culture they live in is not their own. If autistics could live in a world populated only with others like them, their culture would be impoverished in comparison. To be autistic is to be a stranger in a strange land. The child psychologists Penelope Vinden and Janet Astington explore what it may be like to be autistic in a nonautistic world:

'Imagine yourself in this kind of situation—it is actually what very well might happen were you to be dropped into a small village in a Papua New Guinean or Brazilian rainforest. For a while, perhaps for a long while, you would experience what has been called 'culture shock'. You might react in various ways. If you were brave, you might throw yourself into the culture, imitating everything you saw, repeating words that were said to you without understanding, painfully aware that you weren't doing or saying the right thing at the right time, but unable to figure out why. If you were overwhelmed, you might retreat to your hut, and take comfort in your surroundings, trying to make them as familiar and as stable as possible, taking excessive care to keep that little space 'yours' and becoming frustrated and angry with the intrusion of the curious villagers. The villagers, on the other hand, would no doubt regard you as an oddity, and might tolerate you as a somewhat strange child, laugh at you, and, as their initial curiosity wore off, ignore you.' **[Vinden & Astington, 'Culture and understanding other minds' in <u>Understanding Other Minds: Perspectives from Developmental Cognitive</u>**

Neuroscience second ed, Simon Baron-Cohen, Helen Tager-Flusberg, & Donald Cohen, eds, pg 503-519]

Vinden and Astington bring up another way to look at autistic culture:

'People with autism are in some senses individuals without a culture, since culture by its very nature is dialogic. They are persons within a culture, persons whom we enculturate, but who are limited in their ability to reciprocate...we complete ourselves through culture, that there is a gap between what our body tells us and what we need to know in order to function, and culture fills that gap. Individuals with autism are trying to live and function with that void still unfilled.

'However, to say that the child with autism is without a culture is to return again to a view of culture as something we have and get, rather than something which pervades our interactions. Perhaps a better way to view what we interpret as isolation, distance, and aloneness is rather that the autistic child is a culture unto himself or herself. The dialogue, the meaning-making, is taking place within the self, as the self attends to detail, to pattern, to order.' [pg 516]

From the 'neurotypical' perspective, the autistic person's lack of culture must seem tragic, because they miss out on so much that being a part of culture offers. But I think the real tragedy is when they are made to feel as if they are missing out. In contrast, when they are left alone and shielded from the chaotic radiation of the neurotypical 'memeplex', autistic individuals usually are quite content to think and play and live in their own special and cozy little worlds oblivious to the memetic mayhem all around them. They are quite self-sufficient because they are more self than self.

Living within autistic culture has its benefits. Their memes are generally simple and harmless, of the mutualist or commensal type. Thus, there is much less meme-gene, meme-meme, and meme-environment interaction. The genes are able to track the memes on a short leash, rather like the state of nature and culture at the time of the early hominids.

It is conceivable that the autistic phenotype may have been more adaptive in our ancestral environment. The ability and propensity to read minds, deceive and detect deception, and constantly formulate devious plans for undermining others would be maladaptive to life in a world where nothing changed and survival depended on social cooperation. To quote Vinden and Astington:

'Children with autism may be especially out of place in our culture, which is so mentally-oriented. While their restricted capacity for language, which is a cornerstone for learning culture, would severely limit their interactions in any society, their inability to relate intersubjectively might have different outcomes depending on the culture in which they grew up. For example, their desire for sameness, for predictability, might not work against them, as it does in our highly disparate society, but in fact might help them to adjust to living in a culture where there is less diversity. An environment where everyone dresses in grass skirts and bark breech clothes, where every day the main meal consists of taro and squash and greens cooked in a bamboo tube, where daily activities are limited to hunting or gardening—in such an environment the autistic child might find less conflict and more security.'

Such environments were undoubtedly common early in human history, and a few 'primitive' cultures are still rather like this today. Some interesting observations have been make in cross-cultural studies conducted on members of the Ifaluk tribe of Micronesia [Lutz, 1985] and the Baining people of Papua New Guinea [Fajans, 1985]. Lutz and Fajans found that the very concept of the existence of other mental states was alien to these people. They had few words for concepts like 'thought' and 'belief' and 'intention' that pertain to others. Yet they are all intelligent, social people with complete and fully functional cultures. They simply interact with others and anticipate their behavior using their general knowledge base and assumptions based on the memory of past social experiences. Perhaps our 'Eurocentric' concept of TOM is not a human universal after all.

It is ever more difficult for autistic people to cope with modern meme-heavy culture. The pace of technological and social change is breathtaking and disorienting even for neurotypicals, and it shows no sign of slowing down. It is thus both ironic and fortunate that, thanks to the increasing degree of professional subspecialization, more professional niches are opening up for the more capable HFA/Asperger syndrome people. Some of them have undoubtedly led happy and fruitful lives as scientists, engineers, librarians, accountants, clerks, and even medical doctors. The trick is to find them early when their brains are still plastic and before the culture shock of the memeosphere causes lasting psychological damage. Once they are diagnosed, autistic children should be encouraged to use their skills and interests to create a meaningful culture of their own, rather than be forced to modify their behavior, and act 'normal' just to fit into someone else's idea of normal culture.

◆   ◆   ◆

Thanksgiving weekend, 2004. I am at the Ocean County Mall in Toms River, New Jersey. I load up my trunk with Christmas gifts and prepare to drive out of the congested parking lot. I turn on the ignition and look in my rear view mirror. There is a pickup truck idling behind me. I honk. Nothing. I honk again. A guy in a lumber-jacket gets out and walks up to me. He is about 6'2", 230 pounds, with a shaved head, goatee, and tattoos on both arms. He puts his beefy hands on my driver's side window. I feel my pulse race.

'Hey! What the hell is wrong with you? Yeah, you!'

I start to stammer. 'I'm trying to get out and you're blocking my way…'

The man's eyes bulge out as he leans in. 'No, no! You're not gettin' out until I get out, you piece of shit!'

I am very angry and scared at the same time. 'I'm going to call the police,' I threaten, not very convincingly.

'Go ahead, call the police. I am the police!'

For a second I believe him before coming to my senses. He can't be a police officer. I start fumbling for my cell phone.

The man backs off with a parting shot: 'Go ahead, you fucking yellow belly Chink!!'

I find myself stunned beyond words. The redneck has driven off, and I forgot to get his license number.

◆   ◆   ◆

January 8, 2006. I am at the Rubin Museum of Art in New York City with my girlfriend, Jessica. The museum houses a fabulous collection of Buddhist art from Central Asia. The patrons are a microcosm of this wonderfully diverse city: blacks, Asians, Latinos, young children, old ladies, men wearing yarmulkes, women in saris. We are all here to learn something of this peaceful and deeply contemplative religion that stresses love for one's fellow creatures. For me, that simple message goes beyond racial, sexual, religious, and cultural tolerance. We should learn to understand and accept neurobiological differences too. Autistics should try to change what they can and accept what they cannot in learning to live by the rules of a neurotypical world. At the same time, neurotypicals should try to understand and appreciate the special needs, talents, and differences of autistic individuals, and treat them with the respect they would give any other. To love and be loved is what the meme of secular humanism is all about.

9

American Seoulman

Kimpo Airport, Seoul, June 1973. The little man is looking back at me over the bathroom sink. His big head turns from side to side trying to see how his eyes move. They stare straight ahead while the head swivels around them. Very interesting. It is as if there is a magnetic force in the glass attracting those eyes.

'Woong Jae ya! Nawa ra! Ballygaja! Bangi dunan da!' ['Woong Jae, come out! Let's go! The plane is going to leave!]

My mother is calling me. I leave the little man in the lavatory mirror behind. I am leaving on an airplane!

The big room is full of chairs nailed to the floor. There are lots of people sitting around with their suitcases and bags. Some are sleeping on their bags, others are leaning on each other. On the ceiling above us, a big black sign keeps changing. Little squares with numbers and letters written on them flip over very quickly. They make little clicking noises as they turn. Outside the window, there are big airplanes slowly moving back and forth. They are screaming at me.

The airplane speeds up very fast. I look out the window and see the runway pass by in a blur. Then something lifts my chair and makes my ears pop. Outside, trees and buildings get smaller and smaller. People and cars become too small to see. We cross a river with little bridges and boats on it. We pass mountains and lakes, fields and forests. We fly through clouds and more clouds. Soon there is no more land below, just tiny white ripples in an endless indigo sea. The sun soon sets behind us and bathes everyone with a soft purple-orange glow. It must be time to go to sleep. I turn to my mother and try to suck her nipple.

I look out the window. The sky is dark. But there are thousands, maybe millions, of tiny flickering lights all around. It looks like someone had turned the stars upside down. We pass an island with lots of pointy buildings. The tallest ones have blinking red lights on top. My ears pop.

The intercom says something in a language I cannot understand: 'Ladies and gentlemen, we are now approaching Kennedy International Airport. We advise all passengers to keep their seatbelts fastened and their seats and seatbacks in upright position for our final descent.'

Then someone translates it into Korean.

Why do the seats have to be upright, I wonder. I ask my mother. She tells me it's to make the plane land better.

In a few minutes it happens. The strange voice comes on again. 'Ladies and gentlemen, welcome to New York City and the United States of America.' I turn to my mother.

'Mara goren ni? Mara goren ni? Uh, uh? Mara goren ni?' ['What did he say? What did he say, huh, huh? What did he say?]

My father answers, 'Mee gook eh wat da!' [We are in America!]

I am running through a long dark hallway holding my mother's hand. It is hard to keep my feet from bumping into each other. I fall on my knees. My mother sighs and carries me on her back for a while. After we get our suitcases at the baggage claim, she starts to cry.

'Wea woolgoo in ni?' ['Why are you crying?'] I ask her. She keeps crying but doesn't answer me. She looks sad. Maybe she wants to go back. 'Woolji ma' ['Don't cry'] I tell her.

◆   ◆   ◆

The Brooklyn Bridge with its spider web cables and pointy arch towers is on the living room wall. Cars and trucks are driving on it. The little people in the cars are all wearing pointy hats and have pointy bird beak mouths.

On the hallway wall are the skyscrapers: the Empire State Building, the Chrysler Building, and the World Trade Center. Big cranes are on top of the twin towers and bulldozers and cement mixing trucks are moving around at the bottom. Little men with pointy hats and bird beaks are operating them. High above in the sky is a jet plane carrying pointy headed people to the airport.

I saw something new on the black and white television today: a moon rocket. It is 363 feet tall and carries three astronauts from Kennedy Space Center to the moon and back. It is coming out of a huge building on a giant tractor. There are lots of little men in pointy hats and bird beaks working on it. I start to draw it on the bedroom wall. My mom and dad buy me magic markers to finish the drawing.

◆   ◆   ◆

It was my first day in kindergarten at PS 156 in Brooklyn. I was already crying uncontrollably as my mother took me into the dark stairwell. The echoes were very loud. It smelled like old paint and cement powder. I was frightened. She released my hand and told me to go ahead. I screamed for her to take me home. But when I looked back, she was gone. Everyone was yelling and running around. I stood alone sobbing for my mother. A group of boys passed me and called me a crybaby. Ms. Ratner told me to be quiet or else she would call my parents. I told her to please call. When she tried to pick me up and put me on her lap I scratched her face. She screamed and ran to the bathroom. Everyone stopped and stared. Did I do something wrong? I was annoyed and frustrated that Ms. Ratner couldn't see my point of view. Why couldn't she understand what I wanted?

◆   ◆   ◆

Autistic Consciousness

In earlier chapters, I made the point that at its core, autism is a low-level problem involving defective genes, proteins, and neurotransmitters. This is in fact where the problem lies. The defect in autism is not in the process of consciousness but in its contents. It is ironic that a low-level defect in gene expression (perhaps involving synaptic plasticity) produces a global defect in conscious content. But we should keep in mind that consciousness is best approached from many different levels of analysis. To truly explain the phenomenon of consciousness, one must be able to skillfully and seamlessly bind all these diverse levels together.

◆   ◆   ◆

We are now finally ready to tackle the issue of consciousness in autism, or, for want of a better term, the autistic soul. One way to do this is to subdivide consciousness into four categories: phenomenal consciousness (P consciousness), verbal access consciousness (A consciousness), self consciousness (S consciousness), and free will (agency). We can 'map' these four types of awareness onto the brains of three groups of people: 'neurotypicals', autistics, and those with Asperger syndrome (AS) or high functioning autism. (**figure 1**)

	P Consciousness	A Consciousness	S Consciousness	Free Will/Agency
Neurotypicals	+++	+++	+++	+++
Autism	++++	+	+	+
Aspergers	++++	++	+++++ ?	++

Figure 1: autistic consciousness

Phenomenal consciousness is mediated by thalamo-cortical loops involving the temporal lobes and the limbic system. I believe that these circuits are overactive in both AS and autism, resulting in things like impulsivity and excessive attention to detail. Therefore, P consciousness (the 'feeling of what happens') is at least as active (if not more so) in the autistic disorders as it is in normal individuals.

Access consciousness depends on an intact abstraction ability and language processing center. These are believed to be located in the right and left frontal lobes respectively. Either one or both of these regions are underactive or defective in the autistic disorders. More severe autistics are likely to have extensive deficits on both sides, leading to severe problems with both language and TOM. AS individuals have relatively intact language areas, but defective abstraction; therefore, their A consciousness is intermediate between neurotypicals and autistics.

Free will or 'agency' is mediated by the anterior cingulate cortex, in conjunction with the other areas of the prefrontal cortex mediating working memory and executive functions. Both AS and autism suffer some decreased function in these areas, although, again, the deficits are more severe in autism than they are in AS.

Finally, self consciousness is mediated by the self referential body map loops within the parietal cortex, possibly involving mirror neurons. There is some controversy as to whether autistic people have more or less self awareness than neurotypicals. Some researchers have speculated that autistic individuals are impaired in S consciousness. [**Toichi, M., et al, (2002) 'A lack of self-consciousness in autism', <u>American Journal of Psychiatry</u>, 159:1422-1424**] [**Ben Shalom, Dorit, (2000) 'Developmental depersonalization: the prefrontal cortex and self-functions in autism', <u>Consciousness and Cognition</u>, 9:457-460**] This is supported by, among other things, their lack of linguistic self-reference. Another argument in this direction is that it takes an ability to understand and acknowledge other people before one can adequately acknowledge oneself (by reference and comparison). This requires an intact theory of mind. I agree that S consciousness may be impaired in the more severe cases of autism. But I have always felt an acute sense of myself as a separate person. I think one possibility is that the higher functioning autistics or those with AS have just enough abstraction ability to transfer their hyperactive P consciousness to an abstract idea of a unified self, but not enough (for practical purposes) to conceive of other conscious selves. I think it is quite possible that some on the autistic spectrum have rather less self consciousness than neurotypicals, which could explain their penchant for and contentment in engaging in repetitive stereotyped activities that seem to 'flow' automatically. But others on the spectrum, like myself, are very different. We can't put ourselves in Ms. Ratner's place and understand why a screaming boy at school must be calmed down. We are more self than self.

◆   ◆   ◆

Thirty two years have come and gone. I learned English, survived my father and the bullies, finished university and medical school, experienced women, discovered Asperger syndrome, wrote a couple of books, and opened my private practice. I have been complimented on my bedside manner. I have also been told that I speak with a flat Midwestern accent. The little boy from Seoul has become as American as apple pie. Yes, I have learned a lot; especially, pretending to be normal. But there is so much else that lies beyond me. I too often miss the big picture, have difficulty with rules and sports, suffer crippling social anxiety, act on impulse, lose my temper, talk in monologues and tangents, and think and act in disorganized ways. I push myself now and then, and the results are sometimes quite remarkable. But this is still short of a cure. The little autistic boy from Seoul will always be a part of me.

APPENDIX A

Genetics

DNA

All living things on earth are replicators. That is, they are entities with finite life spans that carry the instructions for the near faithful reproduction of their own kind. In this sense, they are technically immortal. Not all replicators are what we would consider 'alive'. Crystals of ice, viruses (of both computer and biological varieties), and even ideas and social customs ('memes' discussed in chapter 8) have the ability to replicate. Indeed, there is some controversy as to the precise definition of 'life' itself. But by convention, living organisms are defined as carbon-based entities that replicate freely (as opposed to doing so by hijacking the biochemical machinery of other organisms as viruses do) through the duplication of genetic code.

'Genetic code' refers to a series of chemical components strung out in a linear structure whose **sequence carries a specific meaning**. Unlike random arrangements of particles in ice crystals, the genetic code is nonrandom, and contains information. There are two types of code: RNA and DNA. RNA is a single stranded molecule found in every living organism. It consists of a compound, called a base, which comes in four flavors: adenine, guanine, cytosine, and uracil, abbreviated A, G, C, and U, respectively. Each base is attached to a sugar molecule, ribose. This structure, called a nucleoside, can form long chains by binding to other nucleosides via phosphate groups. The resulting polymer is called ribonucleic acid, or RNA. An RNA molecule is thus simply a long sequence of nucleosides whose sequence determines its character. RNA has some intrinsic ability to catalyze chemical reactions (called 'enzymatic activity'), and with a little help from some other compounds, is able to replicate itself. For this reason, it is now believed that RNA was at the core of the very first living things on earth some three and a half to four billion years ago. A few simple (and not so simple) viruses, such as HIV, use RNA as their primary means of replication. But in all other organisms, this crucial role has been co-opted by a related but more complex molecule called DNA.

Deoxyribonucleic acid, whose structure was first announced by Sir Francis Crick and James Watson in April 1953 from Cambridge, England, is chemically somewhat different from RNA. It too consists of four different bases: A, G, C, and, instead of uracil, thiamine (T). The bases are linked to a slightly different sugar, deoxyribose. Like RNA, these 'deoxyribonucleosides' are then linked to each other via phosphate groups to form long chains. But the beauty of DNA is that its bases also readily form weaker bonds (called hydrogen bonds) with their counterparts on another chain. Hydrogen bonds, incidentally, are also what keep molecules of water together at its surface, producing its uniquely strong 'surface tension', a crucially important property which has allowed life to arise in the early oceans. In DNA, the base A forms hydrogen bonds with T (or U in RNA), and G bonds with C. Thus a strand of nucleic acid can bind with a 'complementary strand' running in the opposite direction to produce the famous 'double helix' structure.

In both RNA and DNA, the nucleotide base sequence determines its meaning. DNA can replicate itself, and pass on an identical genetic code to the offspring. It can also be used as a 'template' to make RNA, which itself is then used as a second template to make proteins. Proteins are the bread and butter of all life on earth. Virtually all enzymes, most hormones, and many of the structural components of cells are proteins. And all proteins are made from the instructions coded within DNA and its intermediary, messenger RNA. This has come to be known as the 'central dogma of molecular biology'.

DNA and RNA contain the information within them to make the proteins of all life on earth. But how? The remarkable series of discoveries from the elucidation of the double helix in 1953 to the 'cracking' of the genetic code in the late 1960s are rightfully regarded as the greatest collective advance in the life sciences since Darwin's discovery of natural selection. Its impact on biology was profound; it is comparable, and in many ways complementary, to the great revolution in modern physics at the start of the century. A roll call of brilliant men like Linus Pauling, William Bragg, Francis Crick, Sydney Brenner, Jacques Monod, Fred Sanger, Marshall Nirenberg, and others, all Nobel laureates, many originally trained in the physical sciences, paved the way for our current deep (yet still partial) understanding of life.

All proteins from antibodies and enzymes to muscle fibrils and collagen are made up of linear polymers of chemicals called amino acids. There are just twenty different amino acids that occur naturally. The particular sequence of amino acids determines how the polymer will fold and arrange itself into a mature protein, which in turn determines its identity and biological function. When it became clear that DNA and RNA formed templates for the building of proteins from individual amino acids, it was logical that the sequence of nucleotide bases would somehow correspond to the sequence of the amino acids in the mature protein. The 'secret' is that this sequence in DNA and RNA forms a code.

There are four different nucleotide bases in DNA and RNA (recall that RNA uses U instead of T) and twenty different amino acids in proteins. Thus the code cannot be a 1 to 1 match. It cannot be a 2 to 1 match either as 4 times 4 equals only 16 different nucleotide combinations. Francis Crick and Sydney Brenner proposed that the simplest explanation would be a genetic code consisting of triplet base combinations (called codons), since 4 times 4 times 4 equals 64 possible combinations, more than enough to account for all the amino acids.

They were right. By 1967, using ingenious biochemical techniques, Marshall Nirenberg and Har Gobind Korhana had cracked the genetic code. Note that several codons can specify the same amino acid and that some of them, such as UAA and UAG, act as 'stop' signals which terminate the transcription of protein synthesis. Although the genetic code is the same in all living things on earth, mitochondria and chloroplasts, the cells' tiny energy generators which have their own DNA and a few genes of their own, have slightly different nucleotide 'spellings' for a few amino acids. This is evidence that these organelles (functioning parts of cells) may have descended from very ancient free-living bacteria which originated before the evolution of the standard genetic code.

Protein synthesis involves two major steps, **transcription** of DNA into RNA, and **translation** of RNA into protein. By way of illustration, let's examine the synthesis of a common protein in mammals, insulin.

Insulin is an extremely important hormone produced by the pancreas that is involved in metabolism and regulation of blood sugar. Defective or insufficient insulin activity causes diabetes, the most common metabolic disease in the world after malnutrition. The structure and amino acid sequence of the insulin protein was first described by the English biochemist Fred Sanger in 1953, that same annus mirabilis in which the fading but still glorious British nation in a brief shining moment of brilliance discovered DNA, conquered the world's highest mountain, and crowned the second Elizabeth. Sanger won a Nobel Prize for his discovery of a method of determining the sequence of amino acids in proteins. He went on to win another Nobel two decades later for discovering a way to find the sequence of nucleotides in DNA. Insulin consists of two polypeptide chains of 30 and 21 amino acids long connected to each other by sulfide bonds. The gene for human insulin (which codes for the insulin protein) is located on chromosome 7. The gene is 1.4 kilobases (1400 base pairs) long, which is actually quite small by gene standards, and only a small fraction of this codes for the final insulin protein. The coding regions are called 'exons'. There are three exons on the insulin gene separated by two non-coding regions, called 'introns'. The gene also contains a 'promoter' region, which is located just before the first exon, and defines the starting point of transcription, and several 'enhancer' sites some distance away.

Enhancers are specialized areas of DNA that function as binding sites for certain proteins called 'transcription factors' which then allow up-regulation of gene transcription. The insulin gene is transcribed when complexes of proteins attach to the promoter and enhancer regions of the DNA. An enzyme called RNA polymerase is recruited by the complex and starts to move down the DNA helix, using it as a template for RNA synthesis until it encounters a 'stop codon'. The resulting primary messenger RNA (mRNA) transcript is edited by other enzymes that splice out the introns and connect the exons together. The 'mature' mRNA is then transported out of the nucleus and into the cytoplasm for the next step, protein translation.

Translation of mRNA involves two other types of molecules: 'transferRNA' (tRNA), and ribosomes. There are twenty different types of tRNA, one for each amino acid. Each specific tRNA is distinguished by its 'anticodon', a triplet base sequence that matches its complementary codon on the mRNA. For example, the mRNA codon for the amino acid methionine is AUG. The corresponding anticodon is CAU (read in the opposite direction). The codon and its anticodon match up like DNA and its complementary mRNA or the two strands of DNA. The tRNA with the CAU anticodon is specific for methionine and, in fact, is bound to it on the opposite end of the molecule. As noted earlier, all amino acids are specified by multiple codons. Proline, for instance, has four: CCU, CCC, CCA, and CCG. The first two nucleotide 'letters' are specific for proline, but the third is variable. As there are 61 codons for the twenty amino acids (64 minus the 3 stop codons), the code is said to be 'degenerate'. The anticodon of a proline tRNA can bind to any of the four mRNA codons so long as the first two letters are correct. The ubiquitous Francis Crick postulated that this occurs because tRNA is allowed to 'wobble' on its mRNA template, allowing for a less than perfect fit.

In order for the proper tRNA to find its codon on the mRNA, a large wheel-like macromolecule, the ribosome, must first attach to the mRNA strand. The ribosome 'scans' the mRNA codon it currently occupies, recruits the proper tRNA, and removes the attached amino acid. It then moves along to the next codon, recruits a second tRNA, removes its amino acid, and attaches it to the first amino acid via a special enzyme, peptidyl transferase. This whole process involves ribosomal translocation along the mRNA until the entire message has been translated into, in the case of insulin, a 110 amino acid long polypeptide called 'preproinsulin'. Incidentally, it is not fully clear if the ribosome actually 'rolls' along the mRNA like a tiny wheel on a rail, or if it 'slides' along like a tiny zipper. There is some evidence that a rotational mechanism is involved, but that the two main parts of the ribosome are rotating relative to each other, rather than relative to the mRNA strand. [See **Lafontaine, D., and Tollervey, 'The function and synthesis of ribosomes' in <u>Nature Reviews Molecular Cell Biology</u>, July 2001 and Frank, J., and Agrawal, 'A ratchet-like inter-subunit reorganization of the ribosome during translocation' (2000) in <u>Nature</u> 406, 318-322**].

Preproinsulin is not yet a functional insulin molecule. It must undergo several further steps called 'post-translational modifications' which involve multiple cleavages and the bonding of two smaller polypeptide components into the mature human insulin molecule.

Genes and Chromosomes

All creatures great and small replicate by duplicating their DNA (or RNA) and passing a copy on to their progeny. This genetic information specifies the set of proteins that determine, to a great extent, the appearance, behavior, and, in short, the identity of the individual bacterium, slime mold, onion, oak, flatworm, fruit fly, or human being. In eukaryotes (creatures with complex cells that protect their DNA in a nucleus), the genetic information in the DNA is organized into genes spread out among a number of chromosomes. A chromosome is simply a single long molecule of DNA, often tens of millions of base pairs long, intimately wrapped around structural proteins called 'histones'. Each chromosome has many, sometimes thousands of genes along its length. Intriguingly, most of the DNA of most species is not part of any gene, although it may be important to regulate gene expression. Some of this 'junk' DNA contains multiple repeated nucleotide motifs, called satellite regions, that vary from two base pairs to thousands of base pairs in length. These sequences may, in turn, be repeated dozens, hundreds, or even thousands of times. In addition, recall that much of the DNA within each gene occurs in introns, which are never translated into protein. This is further evidence of natural selection at work. To invoke the overused computer analogy, proteins (and bodies) are the hardware; DNA and genes are the software. Our DNA is simply a **working copy** of a giant hard drive duplicated in almost every cell of our body. Much of the stuff on the hard drive is old, defective remnants of once useful ancestral information that nature has not yet gotten around to deleting.

The human genome project, completed in 2001, has successfully sequenced just about all three billion base pairs of DNA in all the 30,000+ genes of all 23 human chromosomes (of at least one person, Craig Venter, the CEO of Celera Corportation). Humans have 22 pairs of autosomal chromosomes plus a pair of sex chromosomes. A few special people have more or less: females with Turner's syndrome, which we will discuss later in the chapter, have just one X chromosome, for a total of 45 chromosomes; men with Kleinfelter's syndrome have two X chromosomes in addition to a Y chromosome; Down's syndrome is caused by having three copies of chromosome 21; some unfortunate infants, who usually die shortly after birth, have three copies of chromosomes 13 or 18. Some men are born with two Y-chromosomes and an X, and some women are born with three or even four X chromosomes. None of these latter conditions are believed to be associated with severe physical abnormalities, although there can be some behavioral abnormalities and mild cognitive deficits.

Generally, simple organisms tend to have less DNA and fewer genes than more complex ones, but this is not an absolute finding. The chicken has 39 pairs of chromosomes to our 23 pairs. The onion has about 15 billion base pairs of DNA to our three billion. Some plants are believed to have two or three times as many genes as we do. But ultimately, it is not the amount of DNA or the number of genes or chromosomes that determine the complexity and behavior of different species or of different individuals within those species, but rather, the **regulation of gene expression** that is most important. The simple fruit fly (Drosophilla melanogaster) that my girlfriend studies has almost as many genes as I do, and perhaps sixty percent of those genes have human homologues. Yet I am much smarter than any fly. Biological complexity lies not so much in the complexity or variability of its components, but 'emerges' from the patterns of interactions of these components at successively higher levels.

APPENDIX B

Consciousness

My earliest memory of **being aware of myself** was as a six year old in Mrs Dixon's first grade class. We were sitting on the school bus on our way back from a school trip to New York City. The other kids were jumping around and being rowdy. I remember wondering what caused them to make so much noise. Eventually the proctor stood up and hollered, "everyone hush up or we aren't leaving!" Most of the kids continued yelling. A few started to look around with wide eyes and frenzied head turning saying, "Sssssshhhhhh!!!". Then one of them yelled, "shut up, you guys, shut up, or we don't go home!" I remember thinking to myself that if everyone kept quiet for themselves, then others wouldn't have to make more noise telling everyone else to be quiet. Then I wondered why, if I could be quiet by myself, everyone else couldn't be like me? What was going on in their heads to make them behave in the particular way that they did? This made me imagine all sorts of things. What would it be like to enter someone else's mind? What would it be like to enter our own minds at some point in the past? Do we remain the same person as we get older? Do sensations feel different for different people, or for the same person at different times? These thoughts eventually led to one urgent question for which I have yet to find a satisfactory answer: **of all the people I could have been, why am I me?** This strange thought stayed in my mind and bothered me as the bus finally grew quiet and crossed the Hudson that evening.

I remember quite vividly the experience of seeing the rapidly passing cables of the George Washington Bridge interrupt the slower passing amber tinted clouds of sunset. First, I focused on the cables, keeping them steady as I saw the cars and clouds pass by. Then I focused on clouds, as I saw the cables pass across. My attention shifted from the cables to the clouds, back to cables, back to clouds as the bus sped across the bridge towards the Palisade Cliffs. Then the questions came back, more urgently than before. **I am me!** I am sure of that, but why me?

Will I always be me? How did I come to be the center of my universe? The feeling was at once deeply disturbing and exhilarating.

◆   ◆   ◆

Over the last decade or so, I have kept track of and often reflected on the ups and downs of my life on a kind of 'scorecard'. The scorecard starts in 1980, when my family and I moved from California to New Jersey, and continues through the present day. Each year is divided into thirds (January to April, May to August, September to December). Each trimester is then given a 'plus', 'minus', or 'zero' depending on whether or not I thought it was a good period of my life.

For instance, the first third of 1980 was a 'zero' because I was routinely harassed by my classmates. This negative was offset, however, by the fact that I was popular with the teachers. The positive and negative feelings I had during this period cancelled each other out and the first trimester of 1980 was a 'draw'. The last third of 1980 was a totally unhappy experience because I had just transferred to a rough public school in Neptune, New Jersey. There, I was bullied daily by the other kids. The most popular people were the jocks and the sexually precocious 13-year-old girls with breasts and an ability to flirt. I was totally left out, and I felt that even the teachers at this new school were indifferent to my plight. So the last third of 1980 earns a 'minus'. 1981 to 1983 was generally a period of emotional and physical disaster for me as the sad pattern continued in three different schools. 1981 and 1982 both have three 'minuses'. These were the worst years of my life.

1984 was my first positive year. By this time, harassment was no longer a major problem for me, and academics gave my life some stability. Also, my self-confidence took a boost when I became involved in the speech and debate club. 1984 scored a 'plus 2', and 1985, a maximum 'plus 3'. But 1986 was a 'negative 2' as I faced major academic and social adjustment problems at a tough new school, the Massachusetts Institute of Technology. I managed to overcome some of these problems by 1988, which, ended up as positive as 1985. The early 1990s were moderately bad years for me. I did have a novel and rewarding experience at Oxford University, but it did not live up to my (highly inflated) academic, social, and sexual expectations. Additionally, in 1991, I had social and academic adjustment difficulties in medical school.

I recovered somewhat by the mid 1990s, but had yet to score a major victory on the sexual relationship front. This was finally resolved in 1996. The year began with a false start. I became infatuated with a young Irish nurse, Nora, at the hospital I was working at as a medical resident. We were both attracted to each other, but I was too timid to consummate the relationship before she left to start midwifery studies at the Hammersmith Hospital in London. I followed her to England, only to be rejected. I was depressed for several months.

Around Halloween, I met another Irish nurse, Helen, at Doc Watsons Pub on the Upper East Side of Manhattan. The cute and feisty Helen was my first real sexual experience. The following year, 1997, was by far the best year to that point. The two of us were in love, and we expressed it with a trip to Hong Kong, just in time to see the sun set on the once mighty British Empire. But the relationship couldn't last because my conservative parents refused to speak to me as long as I continued to sleep with this so-called 'white devil'. Simultaneously, Helen became increasingly demanding, threatening to leave unless I married her. Caught in the middle, I suffered a nervous breakdown in 1998. The last year of our relationship was a topsy-turvy affair that alienated us from (mostly her) friends. We broke up in the summer of 1999. I have not seen her since. 1998 and 1999 were still positive years because I did successfully finish my medical residency despite the emotional hell I went through.

The year 2000 was a near-perfect year, and the greatest turning point of my life (so far). In the first trimester, I received my diagnosis of Asperger syndrome. As I described at the end of the last chapter, this was a truly joyous revelation. The second trimester, I had laser eye surgery, literally giving me near perfect vision for the first time in my life. Finally, in the third trimester, I met my second girlfriend, Jessica, through a dating service for Ivy League graduates. Jessica is a molecular biologist working out the genetic details of fruit fly eye development. I found her work quite interesting. She is also extremely intelligent and well educated, with an English accent, all qualities I find very attractive in women. Finally, she comes from a family of prominent psychologists, artists, and social scientists. By the end of 2000, I had decided to write a book about autism. It was the best year of my life.

◆　　◆　　◆

Throughout this book, I have described bits and pieces of what the brain does (perception, motion preparation, emotion, memory, imagination, decision-making, and so on) and what causes it to happen (evolution, genes, neurotransmitters, synaptic plasticity). I have used autism/Asperger syndrome as the starting point for our exploration of mind and brain. Now we come to the majestic and still mysterious goal of our journey; the point where all these disparate elements come together and connect to 'mean' something for each of us: Consciousness.

Consciousness is at once easy to understand, yet fiendishly difficult to explain. It is something all of us instinctively and effortlessly feel throughout our lives. It is also something about which science has had remarkably little to say, until now. Once the purview of philosophers and religious mystics, consciousness is quickly becoming a legitimate target for neuroscientists, computer specialists, and cognitive psychologists. Thanks to them, we are steadily getting a (somewhat) coherent picture of how the brain creates mind.

◆   ◆   ◆

Before I explain how consciousness happens, we need to define a few loose terms. According to the Oxford American Dictionary: [con-scious-ness n. **1** the state of being awake and aware of one's surroundings and identity (lost consciousness during the fight). **2** awareness; perception (had no consciousness of being ridiculed). **3** the totality of a person's thoughts and feelings, or of a class of these (moral consciousness).] When people commonly speak of 'consciousness' they may be referring to one of at least three types of consciousness. The first type is the subjective **feeling** or awareness of something. It is what it feels like **for you** to taste crème brulee, see a red rose, feel poison ivy, or experience your first orgasm. Such feelings can be triggered by virtually any object or event in the outside world, a memory from the past, or even a figment of the imagination. These personal feelings are private by definition, and cannot be shared with anyone else. Psychologists and others have called this aspect of conscious experience 'sentience', 'aneotic awareness', 'qualia', and '**phenomenal consciousness**'. I will use the term '**P consciousness**'.

A second type of consciousness refers to experiences that we can share with others through verbal report. "That crème brulee at La Bernadin was marvelous," "I'll have to remind myself to buy her a dozen red roses for our anniversary," or "that poison ivy episode last summer ruined an otherwise excellent vacation" are statements that reflect feelings, memories, and plans translated into language. The translation makes our private thoughts—P consciousness—more overtly accessible both to ourselves (we silently 'talk' to ourselves) and to others (we tell them what we think). I will call this '**A (access) consciousness**'.

Finally, there is the awareness of oneself as a **unitary being or agent** experiencing the taste of crème brulee, thinking about a dozen red roses, or remembering one's first sexual encounter. This is the experience of ourselves as a fixed reference point to which all memories, emotions, and perceptions occur. It is the source of all of our decisions, beliefs, and actions. This is the '**I**' which endures through time and travels through space, constantly changing in size, appearance, and health, but still the same entitiy that possesses both P and A consciousness. This is self-consciousness or '**S consciousness**'.

I propose that the three varieties of consciousness are actually distinct entities mediated by different areas of the brain. Moreover, in certain special cases such as autism or schizophrenia, they are dissociable. Because the mind is what the brain does, we need to examine closely those parts of the brain responsible for consciousness (what neuroscientists call the 'neural correlates of consciousness, or NCC for short). But before we zoom in on the physical brain, it is wise to get an overall schematic picture of how the mind constructs consciousness.

A useful way to understand how consciousness works is to take one aspect such as vision or hearing and examine the process from bottom (the eye or ear) to top (visual or auditory awareness). I will utilize this approach by taking a real life example of sensory input and constructing a 'model' of P consciousness that is applicable to all aspects of sentient awareness (perception, memory, emotion, imagination).

◆   ◆   ◆

It was the summer of 1988. I was sitting in a stiflingly stuffy dorm room on Memorial Drive taking a break from studying for my medical school entrance exams. I was listening to a new CD I had just bought—by the heavy metal rock band, Kingdom Come. On track # 9 was the song 'Loving You'. The opening guitar chords resonated in my mind. There was something about it that reminded me of a place I had never been—a beach in Southern California. I skipped the CD back to the beginning of that riff and listened again. The feeling evolved into an almost vivid visual illusion. I pictured myself walking along a boardwalk gazing out at the ocean; it was a hot morning and the horizon was blurred in a hazy chiaroscuro. The chords sounded distant and distorted, as if I were listening to a weak radio signal. I pictured myself languidly walking along that sandy landscape of the mind. What is going on here?

Let's analyze this from bottom to top. First, the train of sound waves from the CD speakers stimulates my eardrums. This causes the three tiny bones of the middle ear to move and create minuscule vibrations in the fluid inside the snail-like coils of the cochlea or inner ear. The vibrations stimulate hair like projections on the specialized nerve cells lining the inner ear (called 'hair cells'). The hair cells are arranged along the length of the cochlea; those at one end have 'long hair' and are stimulated best by low frequency sounds. Those at the other end have 'short hair' and respond best to high frequency sounds. The inverse relationship between length and frequency is a physical principle. It explains, among other things, the changing pitch of musical instruments.

The mechanical stimulation of the hair cells causes a biochemical reaction that opens tiny pores on the nerve cell membrane. This allows positively charged calcium and potassium ions to enter the cell. The influx of these ions changes the electrical charge along the axons of the hair cells, depolarizing them.

Thousands of stimulated neurons responding to a certain sound send axon fibers into the auditory nerve. From there, the signal passes through the brain stem on the way first to the medial geniculate nucleus of the thalamus, and then to the primary auditory cortex of the brain. Like the visual system with its retino topic organization, the auditory processing surface of the thalamus and the cerebral cortex are arranged in a precise spatial order. The neurons representing high frequency sounds are located towards the front and those representing low frequency sounds are located in the rear of both the thalamus and the auditory cortex. This is called 'tonotopic representation'.

Surrounding the primary auditory areas in the temporal lobes are the supplementary and higher order auditory association areas of the brain. There are separate (and more complex) tonotopic maps in these areas. The sounds of the guitar chords are represented and processed here. From the ear to the auditory cortex, sound waves are transduced into electrochemical signals and bundles of action potential spikes whose patterns are represented in almost picture-like 'auditory maps' on the surface of the brain. But none of these 'musical pictures' are in consciousness...yet. The same is true for all the other sensory modalities: vision, taste, smell, touch, pain. To become conscious of these things, we need to go even further up (**figure 1**).

◆   ◆   ◆

The neural codes representing the activity from the optic, olfactory, gustatory, auditory and somatosensory channels are displayed on nerve cell bodies on the outermost millimeter or so of the surface of their respective cortical areas. These are the raw, minimally processed representations of perception flashing on the surface of the brain like some bizarre organic billboard. What happens next? These dynamic 'pictures' on the brain are transformed, edited, and bound together with other incoming pictures.

Simultaneously, there are neural representations of the guitar chords from Kingdom Come, the heat on my skin, the sweat on my forehead, and the view outside my window (the high rise dorms of Boston University just across the Charles River) created while I am sitting listening to the music. They are somehow blended together to create a multimodal 'picture' in the higher-order association cortices. There are many such association areas, corresponding to different sensory modalities. The higher visual areas alone take up much more space than the primary visual cortex. In certain lower primates, half the cortex is devoted to vision. These higher level pictures are created by dynamic ensembles of cortical neurons that continuously cycle back and forth from the thalamus. Each round of activity through these so-called 'cortico-thalamic circuits' creates a slightly different ensemble—a slightly different picture on the brain. The thalamus is both continuously projecting new sensory information from the outside world into the cortex ('bottom-up') and recycling older patterns back to the cortex.

We can use the analogy of consciousness as something like a movie, continuously projecting ever-changing pictures as part of a 'coherent story'. Most neuroscientists and cognitive psychologists are adamant that there is no internal 'viewer' actually sitting somewhere in the brain and watching a movie. Of course, they are right, at least in the conventional sense. I will return to this important concept of the homunculus, or 'ghost in the machine' later. But for now, let's continue with this useful analogy.

Reciprocal thalamo-cortical loops allow cortical patterns representing 'things out there' (sights, sounds, smells, feelings, thoughts) to reverberate temporally, and connect with similarly reverberating representations of other 'things out there'. The thalamus is the engine or motor powering a sort of never-ending movie projector whose images play on the surface of the cerebral cortex from before birth to death. Each time a cortical pattern cycles through the thalamus (over several milliseconds), it is subtly transformed. The integration of thousands of these transformations in time produces the illusion of motion. It is similar to the illusion created by the rapid serial succession of static images in a motion picture reel, each frame slightly different from the last.

If the thalamus is the motor of this motion picture, the higher order association cortices are the screens upon which this movie is projected. The thalamo-cortical loops blend the input from various sensory modalities, but not in any haphazard way. Just as you would expect any good movie theater to project pictures that are clear and sharply focused with sound tracks that match in synch, so it is with the cinema of the mind. I hear guitar chords, feel prickly heat, and see the ripples of water in the Charles River all at once, but I certainly don't 'hear' the heat, 'see' the chords, and 'feel' the river. The input is blended, but also appropriately segregated.

There are some people afflicted with a curious neurological disorder called synesthesia. They actually (not just metaphorically) confuse and experience one sensory modality for another. Some of them become artists who have the gift of being able to 'see' the 'colors of music'. Others are musicians who can 'hear' the 'sounds' of a perfume or a landscape. Some cases of synesthesia are believed to be caused by genetic defects. Other cases may result from traumatic insults to parts of the association cortex or their connections that are important in the proper segregation of different perceptual channels. Instead of enjoying a coherent and sharply focused multifaceted world, synesthetics experience a disconcerting (and perhaps sometimes inspiring) sensory fuzziness. There are reports that some cases of autism are linked to synesthesia.

Finally, V.S. Ramachandran has made a marvelous set of discoveries in amputees. He found that these people can often 'feel' their severed hand or leg in other parts of their bodies, such as their trunk or even genitalia. These sensations range from mild tingling to excruciating 'phantom pain' to perhaps something approaching an orgasm. The sensations that these people 'feel' as being in their nonexistent limb are caused not by something inside the injured stump, but by a 'remapping' of the cortical areas which once received fibers from it. Because of cortical plasticity, denervated areas are taken over by surrounding regions that receive healthy, active input. Note that in the sensory homunculus, the representation of the trunk is adjacent to that of the arm, while the genitalia are next to that of the leg. This explains why gentle stimulation of the trunk (or genitals) can sometimes elicit intense sensations 'in' the lost hand (or foot). Ramachandran half seriously speculates that this may explain the basis for the 'foot fetish'.

◆   ◆   ◆

Consciousness is not an all or nothing proposition. It is not like an action potential or a binary switch. Rather it is more like a sunrise, slowly dawning over a dark landscape of oblivion. There are grades and shades of awareness and self-awareness. The lowest levels (P consciousness) are supported by nerve fibers originating from special neurons deep in the brainstem. These fibers form extensive synapses with the thalamus, the cortex, and with other subcortical areas such as the amygdala, the hippocampus, and the basal ganglia. Together, they form the 'reticular activating system' (RAS). The RAS modulates the activity of the thalamus and its circuits with the cortex. Damage to the brainstem regions housing the RAS (either deliberately inflicted on laboratory animals or resulting from strokes or motorcycle accidents in humans) causes loss of consciousness and coma. Thus the modulation of cortico-thalamic activity by the RAS is necessary for awareness, but by itself it is insufficient to produce consciousness.

Cortical reentry circuits (under RAS modulation) bind lower level sensory 'pictures' into much more elaborate multimodal patterns. But this is only a small part of P consciousness. Recall that in my Kingdom Come reverie, I imagined that I was walking along a beach in Southern California on a hot, hazy morning. How does a real time awareness of tone sequences, heat on my skin, and a view outside my window cause me to imagine something so totally different? The answer is that the visual, auditory, and somatosensory association areas, which are located primarily on the posterior or back parts of the brain are also connected up to four quite different systems: the motor system, the limbic system, the memory systems, and the prefrontal cortex. Each of these systems is a largely self contained module (with submodules and sub-submodules) but they are nonetheless heavily connected with one another and with the thalamus and the RAS in both input and output directions. Perhaps the sensory input I received evoked a memory or an emotional state that in turn was associated with Southern California on a different occasion. Recall that we did discuss these systems in some depth in earlier chapters: the motor system (chapter 3), the limbic system (chapter 4), the memory systems (chapter 6), and the executive control systems (chapter 5). I will offer a brief review here in hopes of elucidating the various aspects of the conscious experience.

◆   ◆   ◆

Motor control, unlike sensory perception, can be thought of as being organized in a top down fashion. Note that I'm using the terms 'top down' and 'bottom up' in a different, though somewhat analogous sense than I did in chapter 8. Here, top down means that a pattern of neural organization is driven by a higher, more abstract level of representation. Bottom up means that more basic elements come together to create a higher level of representation. Just as perception 'flows' from the bottom up (for example, in terms of vision: light>>>retina>>>optic nerve>>>lateral geniculate nucleus of the thalamus>>>primary visual cortex>>>dorsal and ventral streams), motion flows from the top down (supplementary and association motor cortex>>>basal ganglia and cerebellum>>>primary motor cortex>>>spinal cord>>>peripheral motor nerves>>>muscles>>>movement). See **figure 2**.

There is a point on this grossly simplified schematic where the highest, most abstract representations of perception are translated into the highest, most abstract representations of movement. This occurs in the frontal and parietal cortices. These regions represent information beyond simple modalities like motion, color, pain, temperature, and the coding of individual muscle movements. Rather, neurons here code for complex conjunctions of properties integrated across all the senses. The **perception** of sound and heat and vision becomes the **conception** of being in a stuffy MIT dorm room listening to Kingdom Come while gazing out across the Charles River. They also code for a sense of one's body as the reference point for these sensory perceptions.

This last point is very important. In chapter 3, we discussed the concept of reference frames in terms of perception and action. Imagine playing a competitive game of tennis. You find yourself lunging forward to hit a backhand volley. There are several things going on here, each of which your brain must track simultaneously from different frames of reference. First, your legs are moving relative to the court. Second, your arm is moving relative to your body and also to the ball, which, in turn, is moving relative to the court. Third, your eyes are tracking the ball relative to the three dimensional space of the court, the motion of your head, and the motion of your body, all of which are moving relative to one another. How can you keep all of these calculations up to date well enough to return the ball (not to mention handle the automatic and homeostatic activities like respiration and heartbeat)? It's amazing that we can play sports at all.

It turns out that just as there are somatotopic 'maps' of the body in the sensory and motor cortices, there is also a map of the body in the parietal cortex. This third map represents not sensations or movements of the body, but rather the 'knowledge' of one's body as the effector of action. When we decide to move our left foot, for instance, the neural command signal is given not only to the motor system, but also to the 'motor knowledge center' in the right parietal lobe. Additionally, once we move that foot, there is a feedback signal from the muscle fibers all the way back up to this area.

The parietal lobe's body representation map acts as a kind of 'comparator' (like the servo circuit in a thermostat) that tracks our intentions (whether or not to move a part of our body) with their effects. Without it, we would all be bumbling klutzes, forever under or overshooting our goals. There are other such comparator circuits involved in motor control and coordination, notably in the cerebellum and the basal ganglia, as I described in chapter 3. These subcortical organs are responsible for 'automating' repetitive or overlearned actions such as walking or chewing, thus freeing the cerebral cortex to attend to more pressing or demanding tasks. Damage to these subcortical regions such as from chronic alcoholism or Parkinson's disease causes a loss of automaticity such that fine motor control must be accomplished consciously. This results in slow, clumsy movement. On the other hand, the cortex usually exerts an inhibitory effect on subcortical automaticity centers. If subcortical motor regions are hyperactive or no longer inhibited by the cortex, such as in OCD or Tourettes syndrome, motor and cognitive automaticity becomes uncontrollable.

It is in the parietal lobes that intentions are translated into the plans for action. Not coincidentally, it is also where we find those remarkable mirror neurons that allow us to understand others' intentions. Damage to the parietal lobe (usually on the left side) from a stroke or tumor leads to paralysis of the opposite side of the body, and is sometimes associated with **anosognosia**, a condition where the patient denies that she has a physical deficit. It is not that these people are so depressed or shocked that they don't want to acknowledge or think about their paralysis, but rather, anosognosics actually have **no idea** that there is something physically wrong with their bodies! People with anosognosia can be otherwise intelligent and articulate. They understand what it means for others to be paralyzed as a result of brain damage. But they simply cannot grasp the fact that they are now unable to move one side of their own bodies. It is as if they are neglecting half of their bodies. Damage to the right parietal lobe often results in a separate, but related disorder in which the left half of the visual field is ignored—**hemineglect**.

◆   ◆   ◆

Awareness of our sensory perceptions, our movements, and our bodies is usually accompanied by a certain 'mood'. What we see and do makes us feel happy or sad or fearful or tearful. Likewise, tears and fears color our perceptions and motivate our actions. Emotions are an indelible part of conscious experience. In chapter 4, we found that the limbic system, deep within the temporal lobe, is intimately connected both with perceptual input in the posterior cortex, and with the endocrine and the autonomic nervous systems, through the hypothalamus and the pituitary gland (**figure 3**).

The hypothalamus and pituitary gland control many of our bodies' homeostatic processes such as temperature regulation, sleep patterns, and general metabolic activity. We are not directly conscious or in control of these processes. But the hypothalamus also oversees our appetites for three things of which we are very conscious: food, drink, and sex.

The amygdala, the heart of the limbic system, has extensive connections with the hypothalamus, all the higher order sensory areas of the cerebral cortex (especially the olfactory region), the thalamus, and the parts of the temporal lobe important for long term memory storage. As we saw in chapter 3, the amygdala is where thoughts and perceptions acquire their emotional valence, which helps us make better decisions. The emotional valence is, in turn, kept in check by inhibitory modulation from the PFC. An overactive amygdala (or underactive PFC) that may be caused by temporal lobe epilepsy or massive trauma to the frontal lobe often results in a violent, impulsive, emotionally labile personality. Additionally, these people often find anything and everything to have great emotional significance. Superstition, grandiose delusions, and religious ecstasy are some of the symptoms associated with seizures in the area of the amygdala.

On the other hand, an underactive amygdala (or overactive PFC) tends to dampen emotional valence. There is a very rare and wonderfully bizarre disorder called **Capgras syndrome** (usually caused by a blow to the head) whose sufferers come to believe that the people and places once familiar to them are not real. Capgras syndrome patients are convinced that their spouses, relatives, pets, or furniture are not who or what they seem to be, but were somehow replaced with near perfect duplicates by some diabolical mind. These people are not suffering from any form of perceptual defect, amnesia, or cognitive impairment. They simply don't **feel** as if the people and things around them, especially those with emotional significance, are what they should or used to be. It is believed that Capgras syndrome and the related 'depersonalization disorders' are caused by disconnections in the fibers connecting the amygdala with various parts of the sensory cortex.

When I heard the Kingdom Come song, I felt drowsy but also relaxed and content within myself. It was a very compelling feeling I had at the moment; it is also impossible to accurately and completely describe it in words. The emotional modulation that comes from the limbic system suffuses all our perceptions and forms an ever-present backdrop to our thoughts. It is, in fact, virtually impossible to perceive anything without any sort of underlying feeling about it.

◆   ◆   ◆

The feelings and emotions I got by listening to the song reminded me of a beach in Southern California. I had not been to any such beach, but I could certainly imagine it based upon my memories of other beaches and on my knowledge of the geography and climate of Southern California. Emotions are inexorably linked to memories and imagination. This is not a coincidence. As we saw in chapter 6, experiences are converted into memory through the gateway of the hippocampus. The hippocampus stores the 'engram' or memory trace for a period of time before it is transferred to the cortex (long term memory) via the thalamus **(figure 4)**.

Bilateral damage to the hippocampi causes partial **retrograde amnesia** (the inability to recall information experienced or learned shortly before the injury) and **anterograde amnesia** (the inability to lay down any subsequent memories). Patients with this horrible problem retain their personality and intelligence, but are forever trapped in the 'now'. They are fully conscious of their 'core selves', and of the world, but in a very temporary way. So much of who we are depends upon a feeling of something more permanent than this—it requires autobiographical knowledge. This, in turn, requires a healthy hippocampus. As Antonio Damasio writes in his book, <u>The Feeling of What Happens</u>:

The ever changing self identified by William James is the sense of core self. It is not so much that it changes but rather that it is transient, ephemeral, that it needs to be remade and reborn continuously. The sense of self that appears to remain the same is the autobiographical self, because it is based on a repository of memories for fundamental facts in an individual biography that can be partly reactivated and thus provide continuity and seeming permanence in our lives.

The emotional and memory centers of the brain (the amygdala and hippocampus) are intimately connected. Furthermore, both centers receive sensory input, especially from the olfactory system and the ventral visual stream. This is why memories so often evoke strong emotions and emotions evoke poignant memories. It is also why odors, both fragrant and obnoxious, often provoke vivid and visceral memories.

Recall from chapter 3 that the higher order visual areas are loosely segregated into the 'how' or dorsal pathway and the 'what' or ventral pathway. The dorsal pathway passes through the parietal lobe, tapping the 'body reference maps' located there. This allows people to orient their hand movements relative to other parts of their body, to other objects, and to the world in general. Damage to the dorsal stream causes various degrees of **apraxia** (the inability to perform a skilled, learned, purposeful motor act, such a turning a key in a doorknob). People with apraxia are **conscious of what it is they want to do**, and have no sensory/motor problems preventing them from doing them. They simply can't put the act together.

The ventral visual stream terminates in the temporal lobe. It allows us to identify what an object is based upon prior knowledge or memory of similar objects. Damage to this pathway causes **agnosia** (the inability to recognize objects or shapes). It is interesting that people with agnosia and an otherwise intact dorsal stream are able to perform intricate actions such as threading a needle or buttoning up a shirt, without any idea of what it is they're doing. It is as if a 'zombie' (in the dorsal stream) is performing stunts outside of conscious awareness (in the ventral stream). Findings like these have led Ramachandran to propose that the temporal lobe, with its triple access to memories, emotions, and semantic knowledge, is the seat of consciousness.

◆　　◆　　◆

We have now almost finished constructing the 'model of the mind'. What remains is the integration of the sensory, motor, limbic, and memory modules through time. Although Dan Dennett is adamant that there is no commanding homunculus that puts all these disparate activities together in some central 'Cartesian theater', I believe that a spatial-temporal integration of sorts does occur.

The key players are the thalamus, the PFC, and the anterior cingulate cortex (AC). In chapter 5, I described how the PFC, which has evolved to massive proportions in the human species, contains neural ensembles that code for goal-directed associations of sensory, motor, limbic, and mnemonic information. This is where goals are set, complex action sequences are planned, and decisions are made. Psychologists call these functions 'working memory' and 'executive control'. Damage to the PFC causes defects in working memory and executive control, as well as personality problems such as impulsiveness or indecisiveness **(figure 5)**.

This information is then fed into the AC and the thalamus. The thalamus links the goals with incoming sensory information and outgoing motor commands in a looping triangular circuit. The AC is a kind of coincidence detector that is activated by novel or conflicting input. When aroused, it sends signals into the limbic system, which in turn activates the sympathetic nervous system, heightening our sense of awareness, producing the feeling of 'SURPRISE!', and motivating further action. I believe this is where we get the sensation of 'free will'. Damage to the AC can lead to **akinetic mutism**. People with this extremely odd disorder are not unconscious or comatose, but seem to have a total lack of motivation or desire to do anything purposeful. They will sit around all day, gazing off into oblivion, saying nothing at all. In short, they have lost their free will.

◆   ◆   ◆

Step back and take a look at the model mind we have created **(figure 6)**. I believe that this model, as metaphorical as it is, nonetheless captures something of the essence of the mind—both its modularity and its dynamic interconnections. But where exactly is consciousness, or more specifically, the P, S, and A varieties, within this model?

I believe that consciousness is a continuous stream that links each of these clumsy stick figures of the mind together. Note that my 'mind figures' have three major components: 1) sensory/perceptual input, 2) thalamo-cortical re entry loops, and 3) the memory-limbic system. The metaphorical 'stream of consciousness' traverses not all three, but just the second or intermediate level of the mind.

The English computational neuroscientist David Marr speculated (correctly) that we are aware only of certain levels of visual processing. It is quite obvious that the 'lower levels' (the shadows cast by objects reflected on the curved surface of the retina) are not reported directly to consciousness. Likewise, the high level calculations that produce the illusion of solid three-dimensional objects are also not immediately available to consciousness. Rather, it is something in the middle levels of analysis that we are conscious of.

Inspired by Marr and the eminent British-American psychologist Anne Treisman, the psycholinguist Ray Jackendorff has elaborated the notion of consciousness as an intermediate level process. The sensory input to the brain is initially processed by (relatively) simple detectors that transform the information into ever more elaborate representations. The representations start out with an 'observer' or 'self centered' point of reference, but gradually become abstracted into 'object centered' conceptions. The earliest and latest representations do not enter the domain of conscious awareness. As Stephen Pinker writes in <u>How the Mind Works</u>:

"People are unaware of the lowest levels of sensation. We do not spend our lives in Proustian contemplation of every crumb of the madeleine and every nuance of the decoction of lime flowers. We literally cannot see the lightness of the coal in the sun, the darkness of the snowball inside, the pale green-gray of the 'black' areas on the television screen, or the rubbery parallelograms that a moving square projects on our retinas. What we 'see' is a highly processed product: the surfaces of objects, their intrinsic colors and textures, and their depths, slants, and tilts. In the sound wave arriving at our ears, syllables and words are warped and smeared together, but we don't hear that seamless acoustic ribbon; we 'hear' a chain of well-demarcated words. Our immediate awareness does not exclusively tap the **highest** level of representation, either. The highest levels—the contents of the world, or the gist of a message—tend to stick in long-term memory days and years after an experience, but as the experience is unfolding, we are aware of the sights and sounds. We do not just abstractly think 'Face!' when we see a face; the shadings and contours are available for scrutiny.

The advantages of intermediate-level awareness are not hard to find. Our perception of a constant shape and lightness across changes in viewing conditions tracks the object's inherent properties: the lump of coal itself stays rigid and black as we move around it or raise the lights, and we experience it as looking the same. The lower levels are not needed, and the higher levels are not enough. The raw data and computational steps behind these constancies are sealed off from our awareness, no doubt because they use the eternal laws of optics and neither need advice from, nor have any insights to offer to, the rest of cognition. The products of the computation are released for general consumption well before the identities of objects are established, because we need more than a terse **mise en scene** to make our way around the world. Behavior is a game of inches, and the geometry and composition of surfaces must be available to the decision processes that plan the next step or grasp. Similarly, while we are understanding a sentence there is nothing to be gained in peering all the way down to the hisses and hums of the sound wave; they have to be decoded into syllables before they match up with anything meaningful in the mental dictionary. The speech decoder uses a special key with lifelong validity and should be left to do its job without interference from kibbitzers in the rest of the mind. But as with vision, the rest of the mind cannot be satisfied with only the final product, either—in this case the speaker's gist. The choice of words and the tone of voice carry information that allows us to hear between the lines."

◆   ◆   ◆

The Binding Problem Solved

The model mind makes for a nice story, but does it explain how everything 'comes together'? Perceptions, emotions, body images, and memories are somehow integrated into what feels like a coherent, unified 'self' that makes decisions and strives towards goals. This is how the great molecular biologist and neuroscientist Sir Francis Crick puts it:

"Our experience of perceptual unity thus suggests that the brain in some way binds together, in a mutually coherent way, all those neurons actively responding to different aspects of a perceived object. In other words, if you are currently paying attention to a friend discussing some point with you, neurons that respond to the motion of his face, neurons that respond to its hue, neurons in your auditory cortex that respond to the words coming from his face, and possibly the memory traces associated with knowing whose face it is all have to be 'bound' together, to carry a common label identifying them as neurons that jointly generate the perception of that specific face" [Crick, 1992 The Astonishing Hypothesis]

The binding together of aspects of perception, memory, thought, and conscious awareness itself is, I think, the major problem in neuroscience today. There are actually several 'binding problems', which I will divide into three categories before attacking them separately later in this chapter. The first concerns the binding of different modes of perception of the same object. The second involves the binding of these perceptions together in time to create a 'moving picture'.

Finally, there is the issue of how the moving picture tells a compelling story—the contents of conscious awareness.

◆　　　◆　　　◆

One way of accounting for our perceptions would be to imagine that each and every quantum of memory, emotion, knowledge, and even perception that one could conceivably experience has a corresponding neuron representing it. This has been called the 'grandmother cell theory'—referring to a hypothetical neuron representing your grandmother. But a single neuron is just a stupid on-off switch; it carries only one bit of information. Therefore, you would need neurons to represent all the different aspects of your grandmother. There would be a neuron for 'a happy grandmother' or 'a sad grandmother' or 'a sad grandmother in a red polka dot dress looking forlornly out of the window of an otherwise desolate bus depot as the Milwaukee intercity pulls out into the prairies on a late sunny afternoon' and so on. As you can see, the grandmother cell theory is clearly ridiculous. There are simply not enough neurons in a single brain (or even in all the brains in the universe!) to represent all possible contingencies.

Rather, the brain represents events, thoughts, memories, and percepts using a **combinatorial system.** Combinations of synapses dynamically turning on and off throughout the brain contain **all** the information in our heads, both conscious and unconscious. It is not yet clear how these combinations are programmed. One promising idea, proposed by Eric Kandel among others, is that those special modulatory neurons in the brainstem RAS we discussed earlier form intimate synaptic connections with networks of cortical neurons in areas of the brain which process perceptual input or emotions or memories and so on. When something salient or 'noteworthy' captures our **attention** (discussed later), these modulatory cells fire. Activity in these cells somehow binds together all the cortical activity that was occurring at the same time so that it is coded as referring to a single object. This coincidence detector is likely mediated by NMDA receptors (chapter 8) and by modulatory neurotransmitters such as serotonin and dopamine. Because it allows different sets of neurons, each representing different aspects of a stimulus such as its color and sound, to be only temporarily connected, it is known as 'heterosynaptic plasticity'.

There is growing evidence from several different areas of neuroscience that heterosynaptic plasticity is also the mechanism that binds the contents of perception, memory, and imagination with a sense of 'value', perhaps elevating those contents above a critical threshold of conscious awareness. Emotional valence via the amygdala and motivation-reward signals from the PFC, for example, appear to rely on heterosynaptic plasticity. The neuroscientist Earl Miller writes:

"A major source of reward-related signals may be the dopamine-mediated innervation of the PFC from a group of cells situated in the ventral tegmental area (VTA) of the midbrain…The resulting dopamine influx into the PFC could affect plasticity through several plausible mechanisms. For example, dopamine could augment NMDA receptor mediated glutamatergic transmission…When the dopamine influx reaches the PFC, it could strengthen connections-associative links—between neurons that were activated by the event that preceded it…PFC activity could exert a top-down influence by providing an excitatory signal that biases processing in other brain systems towards task-relevant information." **[Miller, E., 'The prefrontal cortex and cognitive control' <u>Nature Reviews Neuroscience</u> Oct 2000]**

◆   ◆   ◆

Heterosynaptic plasticity and subcortical modulation of cortical processes are certainly important and probably necessary for binding the contents of consciousness together. But I don't think they are sufficient by themselves. This level of integration produces static 'pictures': for example, the sound of birdsong and the feeling of nausea induced by a memory of a sailing trip. But in order to advance to the next frame, these pictures must **move** in time. As I discussed earlier, the illusion of a moving picture occurs through thalamo-cortical loops. Let us examine these loops more closely and see what binds them together. They really are collections of millions of neurons in various areas on the surface of the cerebral cortex (the gray matter) that send millions of axons deep into the brain (the white matter) to synapse with other neurons in the thalamus. The cells on the surface of the cortex, called pyramidal neurons, are linked to each other in specific patterns of excitation and inhibition. When one pyramidal cell is stimulated by its neighbor or through heterosynaptic modulation from a distant subcortical neuron, it activates other cortical cells in its immediate vicinity (its 'receptive field') but inhibits those some distance away.

Groups of pyramidal cells form patterns of 'attractors' like the swells and troughs of ripples in a pond or waves on the sea, constantly activating and inhibiting neighboring groups of neurons. Moreover, these attractors are arranged in **cortical columns** that have an internal organization. In chapter 3, we saw that the primary visual cortex is organized into columns of adjoining cells which each code visual stimuli with a particular orientation. These orientation columns are then further arranged into ocular dominance columns, which depict input from opposite eyes, and hypercolumns, which also process color input. Many other areas of the cortex beyond the visual areas have well defined 'topical' organizations that map to some logical pattern in the outside world, such as the sensory surface of the body, or the knowledge of living things versus inanimate objects. Axons from these cells synapse on thalamic neurons, which also seem to have corresponding maps of various different sensory, motor, or conceptual domains. Sir Francis Crick and his junior partner, Christof Koch, believe that the matching maps on the thalamus and cortex are important in binding the representations of 'things' (in the cortex) with attention (mediated by the thalamus), elevating these representations into conscious awareness. Sensory input from the area being attended to arrives at the thalamus and is transmitted to the corresponding region of the cortex. The cortex then signals back to the thalamus to link this sensory picture with the input being received at the next instant in time. Crick writes:

"...any one cortical area is strongly associated with only one part of the thalamus...Each might have its own characteristic processing time, its own characteristic times for very short term memory, and most important, **its own particular form of representation**; simple features in V1, 2 ½D objects in the next higher cortical areas, and so on. **The character of each type of processing unit would depend on the content and organization of that particular representation.** Each particular thalamic region may handle its own form of attention, possibly by allowing neurons in its member set of cortical areas to talk to neurons in the thalamus, which in turn feed back to them, so that in some way their firing is coordinated. There is also the speculative idea that the thalamo-cortico-thalamic circuit may be intimately involved as a reverberatory circuit for very short term memory." [Crick, 1992]

If this is true, then recurrent thalamo-cortical connections can be thought of as producing a moving picture generated by thalamic activity constantly transforming cortical representations by shifting attention in time.

◆   ◆   ◆

It is amazing that the billions of synapses it presumably takes to build up a representation of some complex concept keep their integrity through repeated trips from the cortex to thalamus and back, all while a blizzard of new information and 'random noise' constantly reset the cortical attractors. It is as if we took a completed 500 piece jigsaw puzzle, threw it off the Grand Canyon, and it landed in the Colorado River below pretty much intact. The answer may involve **synchronicity**.

Activated pyramidal neurons create a field of excitation around them, lowering the threshold at which their neighbors will fire. This facilitates synchronized firing of neighboring neurons. Because these neurons often synapse with cells many millimeters away, large zones of activity can arise. Within these zones, neural activity was found to oscillate at various different frequencies depending on what the individual was experiencing or attending to. Sleep is associated with a neural firing frequency of 12-25 cycles per second. This is known as the 'beta rhythm'. Wakefulness, on the other hand, is associated with the 'gamma rhythm' (25-70 cycles per second). Some neuroscientists have speculated that the gamma rhythm might be the neural correlate of conscious experience. This is supported by the fact that unconscious states such as epilepsy and deep sleep involve a slower 'hypersynchronization' of neural activity. Additionally, altered states of consciousness, such as schizophrenic hallucinations, also seem to be associated with a loss of the gamma oscillation.

Perhaps the synchronization of large ensembles of neurons firing at a certain frequency helps to integrate or 'bind together' the different features of a perceptual experience. The idea of neural oscillations acting as a kind of 'glue' or 'neural correlate of consciousness' (NCC) is appealing at an intuitive level, but there has not yet been much direct evidence to support it in humans and other primates. Of course, these oscillations would not be confined to one set of neurons, but would move around the brain depending on the contents of consciousness.

◆   ◆   ◆

Thus far we have painted a coherent moving picture, actually several simultaneously moving pictures in a 'mental multiplex', (corresponding to distinct perceptions, emotions, memories, and so on) generated throughout the cortex by bottom-up activation from both externally sensed stimuli and internally generated activity. The features of these pictures are bound together perhaps by synchronized oscillations of linked neural ensembles. These pictures are also biased by signals from modulating neurons in the brainstem and the limbic system. They are cycled to and from the thalamus to the cortex, acquiring further bottom-up and top-down modifications along the way. But these motion pictures, as coherent and accurate as they may be, cannot yet be conscious. The model is still incomplete because there is no 'story'. It is simply scintillating patterns of neural activity, nothing more, nothing less, no matter how hard we look. **Yet there is nothing more personally compelling than the existence of our own awareness.** Consciousness implies a certain salience to our experiences, a coherent reference point in terms of subject and object. We are aware of things in space moving in a continuous trajectory through time. Yet through it all, there is a **focus**, a single point of reference around which all our perceptions, motions, emotions, imaginations, and memories revolve.

This is an illusion.

There is no central focus in the brain, what Dan Dennett disparagingly calls a 'Cartesian Theater'. But the way in which the parallel streams of moving pictures shuttle through the thalamus creates a sort of computational bottleneck. Neural processing is like a massive multilane freeway with many parallel lanes of traffic. But as we move in towards the thalamus, those lanes progressively merge until there is only a single stretch through which all traffic must slow down and pass serially. This creates the illusion of a single focus of attention. It is through this focus, this 'spotlight of attention', that a single unified and coherent stream of consciousness flows. It is the serial stream arising out of a massively parallel architecture that gives rise to our often profound feeling of a qualitative salience—this is the essence of P consciousness.

To see how this feeling happens, we would do well to examine a well-studied aspect of consciousness, the phenomenon of visual awareness. This is how Sir Francis Crick puts it:

"A common metaphor is that there is a 'spotlight' of visual attention. Inside the spotlight the information is processed in some special way. This makes us see the attended object or event more accurately and more quickly and also makes it easier to remember. Outside the 'spotlight' the visual information is processed less, or differently, or not at all. The attentional system of the brain moves this hypothetical spotlight rapidly from one place in the visual field to another, just as, on a slower time scale, you move your eyes.

"The spotlight metaphor in its simplest form rather implies that the visual system pays attention to one **place** in the visual field. There is much indirect evidence that it does this. An alternative is that attention is paid, not to a particular place, but to a particular object. If the object moves (the eyes being kept still) then in some cases attention can be shown to be attached to the object rather than staying in one place. At the moment it seems likely that both forms of attention—to a visual object or to a visual place—can occur to some extent." [Crick, 1992, pg. 62]

Visual awareness seems to arise from and depend upon selective attention to the visual field. Without attention, visual processing still occurs from the retina through the LGN of the thalamus, to V1, V2, and beyond. Along the way, there are modules that detect various individual features such as motion, edge orientation, brightness, color, shape from shading, maybe even faces. But none of this parallel processing is 'conscious' because there is no spotlight of attention to tell 'us' what is salient. **Saliency** or the **'quality of experience'** usually involves the **conjunction of features**. Examples include a round ball that has a certain color, shape, and motion, or a particular face that expresses a certain emotion and evokes specific memories. Parallel processors may be good at detecting individual features, but are not particularly good at calculating conjunctions. Only a serial search can do that.

Thus the contents of conscious experience, which are usually conjunctional, must first pass through a serial processor that focuses attention on specific combinations of features. The only processors in the human brain that do this are the re-entrant thalamo-cortical loops. **Parallel processors outside these loops perform pre-attentive, subconscious computations, but serial processors embodied by the loops perform attentive, conscious ones.** Stephen Pinker describes it like this:

"Why is visual computation divided into an unconscious parallel stage and a conscious serial stage? Conjunctions are combinatorial. It would be impossible to sprinkle conjunction detectors at every location in the visual field because there are too many kinds of conjunctions. There are a million visual locations, so the number of processors needed would be a million multiplied by the number of logically possible conjunctions: the number of colors we can discriminate times the number of contours times the number of depths times the number of directions of motion times the number of velocities, and so on, an astronomical number. Parallel, unconscious computation stops after it labels each location with a color, contour, depth, and motion; the combinations then have to be computed, consciously, at one location at a time." [Pinker, 1997, pg 141]

Recent studies have shown that the process of visual awareness is indeed done serially on top of distributed parallel circuits, which would explain the temporal and spatial resource limitations of attentive search tasks. The Anglo-American psychologist Anne Treisman has devoted much of her career to this issue:

"Some features of surfaces and depth can be detected without focusing attention: thus, parallel detection is possible for properties defined by stereoscopic depth, constancy scaling of brightness, and geometric forms seen as three-dimensional objects lit from a particular direction. These properties conjoin many simple retinal cues, in apparent conflict with the claim that serial attention is needed for binding. One resolution is that parallel search reflects not pre-attentive processing but global attention to the scene as a whole, with no individuation of separate elements. A discrepant object is seen as a break in the global structure of the three-dimensional array. Similarly, a face can be seen as a global whole without requiring serial processing of each feature in turn...Conjunctions of parts may create emergent features, such as closure, which also allow parallel detection. Extensive practice may form new conjunction detectors that mediate efficient search, suggesting the possibility that 'grandmother cells' or local 'cell assemblies' may develop to solve the binding problem for familiar sets of items such as letters, or for useful components of more complex shapes...Recent evidence continues to support the idea that binding depends on attention unless one or other of these special conditions holds." [**Treisman, A, 1996, 'The binding problem', <u>Current Opinion in Neurobiology</u>**]

◆   ◆   ◆

What happens outside the spotlight of attention? Parallel processing of sensory, emotional, mnemonic, imaginary, and other mental information undoubtedly occurs, but the results (usually) remain hidden from view. Sometimes, however, these subconscious processes bubble up to the surface. **Blindsight** is one example. People who have suffered strokes to the primary visual cortex are effectively blind over the corresponding field of vision. When shown a stimulus within their 'scotoma' ('blind spot'), they report seeing nothing. Yet when psychologists persistently prompt them to guess where this 'invisible object' is, they often 'guess' correctly more often than chance should allow! It turns out that there are visual pathways from the retina to the superior colliculus, a sort of primitive subcortical brain structure used to orient eye movements. These subcortical pathways bypass the thalamus and are therefore not part of conscious awareness;

yet they can play a part (although small) in guiding our behavior. Incidentally, in some 'lower' animals such as frogs, the collicular visual pathway is the major route from eye to brain. In such creatures, all sight is 'blindsight'.

Alternatively, things that are usually safely within the charmed circle of attention can sometimes fall outside. This occurs in hemispatial neglect (caused by damage to the right parietal lobe) where the left half of the visual field, or sometimes the left half of objects (in either field) are lost from conscious awareness. The right parietal lobe orients attention to a particular frame of reference. Damage here prevents the attentional spotlight from shining on the left visual field.

Balint's syndrome, caused by bilateral damage to the occipital/parietal regions, interferes with the ability to focus attention on more than one object at a time. When such patients are shown an array of red and green circles and asked to name the color, they pick the color of the object upon which they happen to focus attention. The other objects are neglected because the dorsal stream (which locates objects in space) is knocked out.

Even in 'normal' people, most of what we see cannot be inside the attentional spotlight all at once. Attention works serially. We experience, know, and can describe whatever is within the spotlight, but have a very foggy idea about what is not. This explains why the perceptual features of unattended objects (things seen out of the corner of the eye; our recollections of the colors of people's clothing after a cursory glance) often 'blend' erroneously.

The spotlight of attention is constantly moving around, illuminating a much wider scene one stretch at a time. The contents of attention are kept in short term memory and constantly compared with the updated version, as the thalamo-cortical circuit traverses newly re-entering loops. Because we remember the contents of the previously attended location, the ever-fleeting spotlight's trace is experienced not as a static snapshot, but rather as a fully coherent and continuous movie of the world. Moreover, our ideas, knowledge, moods, and other memories color this illusory moving picture from the top down.

If thalamo-cortical loops create the illusion of a moving picture in our center of attention, then the **movement of that center of attention through time produces the illusion of a continuously conscious self.** If, as a result of damage to the thalamus or to the parietal lobe regions responsible for shifting attention from place to place, the spotlight were to stop moving and fixate on one object only, I suspect that our conscious awareness of the contents of the spotlight would be lost.

A good analogy are the rapid eye movements we make while awake. Our vision is concentrated on the 15 or so degrees of space reflected on our foveae. The remainder of the world is either invisible (because our eyes don't see it) or poorly visible (through peripheral vision) or subconsciously visible (through blindsight). Yet we do not experience the world as if looking through a peephole. Each second we are awake and alert our eyes make three to five involuntary movements called **saccades**. Saccades function to stimulate the retinal cells with continuous fresh visual input while simultaneously providing an up to date and useful picture of the visible world. Like most neurons, retinal cells are activated not so much by the absolute magnitude of the perceived stimulus, but rather by its difference from surrounding space or preceding time.

If the eyes failed to saccade, there would be little change in the context of the observed scene and slowly the picture in the mind's eye would fade to black. Saccades produce the illusion of visual continuity by moving the 'spotlight of vision' around to create a seamless ribbon of visual experience. Likewise, attention shifting produces the illusion of subjective continuity by moving the 'spotlight of attention' around to create a seamless ribbon of conscious experience.

What controls the position of the spotlight of attention? Both top down and bottom up processes can contribute. For example, when reading a book your attention may be driven by the words on the page, or consciously shifted by the realization that reading on a boat will make you sick. When, for whatever reason, the balance between bottom-up and top-down factors is upset by a bias for lower level 'capture', there is a tendency for impulsiveness and attention deficit. On the other hand, if the balance is shifted towards the top-down processes, the result is inflexibility and maladaptive adherence to routines. Incidentally, as we saw in chapter 5, both situations are often found in autism.

◆   ◆   ◆

When the spotlight is moving well, smoothly tracking targets as they appear in the outside world, but also under a healthy amount of regulation by higher level executive control, P consciousness arises. This phenomenon is, of course, an emergent property that we 'transfer' to the objects of perception, but does not actually exist within them.

Ever since Descartes sounded the bell with his <u>Meditations</u>, Western philosophy has grappled with the 'mind/body problem'. It is interesting to examine the views of several British Empiricists on this matter. The Englishman, John Locke, a classic dualist, believed that there was an ultimate physical reality, which he called 'primary qualities', underneath our perceptual awareness of it ('secondary qualities'):

"The particular bulk, number, figure and motion of the parts of fire or snow are really in them, whether anyone's senses perceived them or no; and therefore they may be called real qualities, because they really exist in those bodies...Take away the sensation of them; let not the eyes see light or colours, nor the ears hear sounds; let the palate not taste, not the nose smell; and all colours, tastes, odours, and sounds, as they are such particular ideas vanish and cease, and are reduced to their causes, i.e. bulk, figure, and motion of parts." [**Locke, 1689 <u>An Essay Concerning Human Understanding</u>**]

But Locke also believed in a 'spiritual world' just as real as, although in a separate realm from, the physical world. The Irishman Reverend George Berkeley, on the other hand, denied the existence of the material world in favour of an ideal spiritual reality encompassing all things (including our conscious minds). We harbour the illusion of living in a material world, but are actually figments of God's divine if inscrutable imagination.

The Scotsman, David Hume, jettisoned both dualism and idealism altogether, believing that the only reality is the relation of perceptual processes:

"For my part, when I enter most intimately into what I call myself, I always stumble on some particular perception or other, of heat or cold, light or shade, love or hatred, pain or pleasure. I never catch myself at any time without a perception, and never can observe anything but the perception." [**Hume, 1739 <u>A Treatise of Human Nature</u>**]

A slightly later English thinker, David Marr, brought back the dualist view with his theory on vision (Marr, 1982 <u>Vision</u>). Marr argues that there is indeed an objective material world 'out there', which we perceive through our eyes. The initial steps of this sensory transduction create a simple two-dimensional representation in our retina and primary visual cortex in terms of simple features (edges, color, movement, etc.). This representation is then further processed in the dorsal/ventral visual streams to create a more integrated 'two and ½ dimensional sketch' (contour, texture, depth). Finally, even further downstream, our brains implicitly 'fill in' more information to produce a (usually) fairly accurate facsimile of the three dimensional world around us.

Marr's '2 and ½ D sketch' model, which inspired Treisman and Jackendorf's idea of consciousness as an intermediate level process, is appealing not just for its technical rigor, but also for its philosophical stance. What appears at first glance to be a latter day version of dualism (at least with respect to visual consciousness) is actually totally compatible with materialism. The secret ingredient is the concept of 'emergence'. Marr does not need to invoke ghosts in the machine. He did not believe in a 'little man' sitting in the pineal gland (as Descartes proposed) looking at the subject looking at the object (looking at the subject looking at the object…ad infinitum) in some endless regression. But he did neatly differentiate the physical world from our perception of it. The brain **enables emergent phenomena** such as vision, memory, semantic knowledge, and consciousness. However, we should realize, as Berkeley did in the early Eighteenth Century, that 'ultimate reality' is never directly accessible to us (independent of our sensory perceptions of it). As the great pioneering modern physicists Neils Bohr and Werner Heisenberg discovered to their chagrin in the 1920's, the act of sensing or measuring reality will always interfere with the underlying physical state.

◆   ◆   ◆

In retrospect, I think it is quite obvious why and how this roving serial processor, the 'spotlight of attention', evolved. It was naturally selected for certain survival advantages. Simple organisms such as bacteria manage quite well with a rather limited repertoire of behavioral responses programmed into their genes. They have little need for complex information or long term planning of coordinated movements; hence they have no sensory transduction systems. Such organisms make do with self-contained molecular signaling pathways. As organisms became multicellular, there was selective pressure for more efficient spatial and temporal data integration. Communication among cells within different parts of the organism and between different organisms became important for general homeostasis, self-defense, the acquisition of food and sexual partners. Higher organisms evolved strategies to deal with more complex problems through both preset instructions specified by their genes, and more flexible 'learning mechanisms'. Certain cells became specialized for higher level information processing leading to the evolution of the endocrine, immune, and nervous systems.

The evolution of the nervous system was further enhanced by anatomical and physiological developments presumably allowing neural representations of the outside world. No doubt, consciousness could not have evolved without the necessary increases in brain size and complexity, especially (in humans) in the prefrontal cortex, the seat of working memory and general cognitive control. There are many popular 'explanations' for the evolutionary increase in human brain size. Among them are the need to coordinate the opposable thumb, process symbolic language, and decipher and monitor increasingly intricate social behavior. It is difficult to determine if these or other factors were causal or consequential to brain evolution. But whatever happened, the result was this: the ability to focus and maintain attention on a certain aspect of the environment in relation to one's own body (or part of the body), create a moving 'egocentric picture', take this motion picture, compare it to other motion pictures stored in or generated by other parts of the brain, enhance the comparison, and decide on a course of action. All this in order to deal with constantly shifting and increasingly complex environmental demands in real time. In addition, as certain individuals acquired random mutations in genes which enhanced certain information processing abilities, and as these mutants spread throughout the population, there was mounting pressure for the evolution of countermeasures to deal with or defend against such individuals.

Perhaps the need for better executive control, self-monitoring, central coherence, impulse inhibition, deception of and deception detection in others all contributed to the emergence of brains capable of salient thought and self reflection, which we would call 'consciousness'.

◆   ◆   ◆

We have now almost completed the link between attention and consciousness. Re-entrant thalamo-cortical loops are (almost) always necessary for conscious awareness. Without attention, there can be no consciousness. But is attention sufficient? Consciousness is not a thing but a process. It involves the continuous summation of the attentional trajectory as the serial spotlight moves from place to place over time, illuminating different people, places, and things, body states, emotional states, memories, ideas, and so on. This can be expressed in mathematical terms.

Attention (**A**) refers to the sum of the contents of all synchronized thalamo-cortical activity (**TC**) at a given time that **then subsequently** re-enter through the prefrontal cortex (the act of being 'attended to'):

$$A = \int TCdt$$

Consciousness involves a **second integration process** that sums the contents of attention over time. In other words, consciousness is simply the serial summation, or integration, of whatever 'pictures' happen to be in the spotlight of attention as it traverses parallel streams of re-entrant neural activity:

$$C = \iint TCdt = \int Adt$$

When attention is directed at something other than oneself, the resulting awareness is **P consciousness**. But the spotlight can just as easily be pointed at oneself (via the body reference maps in the parietal lobe) or some aspect thereof (my right arm, my heart, my love for you, etc.). This results in **S consciousness**. The final type of conscious awareness, **A consciousness**, refers to verbal access. We'll get to that later.

There is an emerging theme here: consciousness may be an integration of attention, but attention is directed at **distinct targets**. The molecular biologist-turned neuroscientist Gerald Edelman has incorporated this idea into his 'Dynamic Core Hypothesis' of consciousness. He writes:

1. A group of neurons can contribute directly to conscious experience only if it is part of a distributed functional cluster that, through re-entrant interactions in the thalamocortical system, achieves high integration in hundreds of milliseconds.

2. To sustain conscious experience, it is essential that this functional cluster be highly differentiated, as indicated by high values of complexity.

"We call such a cluster of neuronal groups that are strongly interacting among themselves and that have distinct functional borders with the rest of the brain at the time scale of fractions of a second a 'dynamic core', to emphasize both its integration and its constantly changing composition. A dynamic core is therefore a process, not a thing or a place, and it is defined in terms of neural interactions, rather than in terms of specific neural location, connectivity, or activity. Although a dynamic core will have a spatial extension, it is, in general, spatially distributed, as well as changing in composition, and thus cannot be localized to a single place in the brain. Furthermore, even if a functional cluster with such properties is identified, we predict that it will be associated with conscious experience only if the reentrant interactions within it are sufficiently differentiated, as indicated by its complexity.

"…What is the complexity of the dynamic core itself, and can this complexity be correlated with the ability to differentiate, which is such a fundamental property of conscious experience? A strong prediction based on our hypothesis, is that the complexity of the dynamic core should correlate with a person's conscious state. For example, we predict that neural complexity should be much higher during waking and REM sleep, when we are conscious, than during the deep stages of slow-wave sleep, when we are not. It should be extremely low during the unconscious spells accompanying epileptic seizures, despite the overall increase in brain activity. We also predict that neural processes underlying automatic behavior, no matter how sophisticated, should have lower complexity than neural processes underlying consciously controlled behaviors. Finally, a systematic increase in the complexity of coherent neural processes is expected to accompany cognitive development, in parallel with an extraordinary increase in discriminatory abilities." **[Edelman, G. & G. Tononi, 2000, <u>A Universe of Consciousness</u> pg 144, 154]**

This theme is echoed by V. S. Ramachandran. At the close of his book, <u>Phantoms in the Brain</u>, he proposes three principles of qualia (P consciousness). First, the input must be 'irrevocable'. In other words, the perceptions, feelings, memories, ideas, and so on, whether reflecting something real 'out there' or just 'in here', must activate the early sensory cortical areas producing 'a stable, finite, irrevocable representation in your short term memory as a starting point.' This is analogous to Edelman's first postulate of consciousness: the integration of thalamo-cortical loops.

Second, on the output side, qualia-laden sensations 'afford the luxury of choice'. This means that the experience of P consciousness does not constrain the individual/organism to one mode of response. She or he can do or think whatever she/he wants with it. This is a bit misleading, since our response is **always** constrained by the laws of probability, thermodynamics, biochemistry, and so forth, leaving free will as an illusory, emergent property. But the point is that the possible number of potential outcomes from a single integrated input is rather large, perhaps infinite. This corresponds to Edelman's second postulate: the 'functional cluster' (the attentional loop) is complex and differentiated.

Ramachandran's third principle refers to the ability to hold a subjective representation in short term memory, allowing it to be modified and using it to build up further representations. To quote Ramachandran:

"...for qualia to exist, you need potentially infinite implications but a stable, finite, irrevocable representation in your short-term memory as a starting point. But if the starting point is revocable, then the representation will not have strong, vivid qualia. Good examples of the latter are a cat that you 'infer' under the sofa when you only see its tail sticking out, or your ability to imagine that there is a monkey sitting on that chair. These do not have strong qualia, for good reason, because if they did you would confuse them with real objects and wouldn't be able to survive long, given the way your cognitive system is structured. I repeat what Shakespeare said: 'You cannot cloy the hungry edge of appetite by bare imagination of a feast.' Very fortunate, for otherwise you wouldn't eat; you would just generate the qualia associated with satiety in your head. In a similar vein, any creature that simply imagines having orgasms is unlikely to pass on its genes to the next generation.

"Why don't these faint, internally generated images (the cat under the couch, the monkey in the chair) or beliefs, for that matter, have strong qualia? Imagine how confusing the world would be if they did. Actual perceptions need to have vivid, subjective qualia because they are driving decisions and you cannot afford to hesitate. Beliefs and internal images, on the other hand, should not be qualia-laden because they need to be tentative and revocable. So you believe-and you can imagine—that under the table there is a cat because you see a tail sticking out. But there **could** be a pig under the table with a transplanted cat's tail. You must be willing to entertain that hypothesis, however implausible, because every now and then you might be surprised.

"What is the functional or computational advantage to making qualia irrevocable? One answer is stability. If you constantly changed your mind about qualia, the number of potential outcomes (or 'outputs') would be infinite; nothing would constrain your behavior. At some point you need to say 'this is it' and plant a flag on it, and it's the planting of the flag that we call qualia. The perceptual system follows a rationale something like this: Given the available information, it is 90 percent certain that what you are seeing is yellow (or a dog or pain or whatever). Therefore, for the sake of argument, I'll assume that it **is** yellow and act accordingly, because if I keep saying, 'Maybe it's not yellow', I won't be able to take the next step of choosing an appropriate course of action or thought. In other words, if I treated perceptions as beliefs, I would be blind (as well as being paralyzed with indecision). Qualia are irrevocable **in order to eliminate hesitation and to confer certainty** to decisions. And this, in turn, may depend on which particular neurons are firing, how strongly they're firing and what structures they project to." [Ramachandran 1998 pg 241-242]

◆   ◆   ◆

We have now finished sculpting the **process**, and are now ready to fill in the **contents of consciousness**. Before we move on, let's step back and behold the sight of this magnificent edifice, my metaphorical model mind in all its majesty **(figure 6)**.

Self Consciousness

At the core of the model is the loop created by the reentrant spotlight of attention. As the parallel processors in various parts of the brain do their thing, the spotlight etches out a single 'stream of consciousness.' The focus of attention shifts constantly; the intensity and size of the spotlight itself will change depending on motivation, energy, activity, distractions, etc. But this stream of consciousness is what creates the sense of self. Douglas Hofstadter points out that self-awareness can create consciousness in a mathematical system as well as a brain:

"...the Godelian strange loop that arises in formal systems in mathematics (i.e. collections of rules for churning out an endless series of mathematical truths solely by mechanical symbol-shunting without any regard to meanings or ideas hidden in the shapes being manipulated) is a loop that allows such a system to 'perceive itself', to talk about itself, to become 'self-aware', and in a sense it would not be going too far to say that by virtue of having such a loop a formal system **acquires a self**.

"When and only when such a loop arises in a brain or in any other substrate, is a **person**—a unique new 'I' brought into being. Moreover, the more self referentially rich such a loop is, the more conscious is the self to which it gives rise. Yes, shocking though this might sound, consciousness is not an on/off phenomenon, but admits of degrees, grades, shades. Or, to put it more bluntly, there are bigger souls and smaller souls." [**Hofstadter, <u>Godel, Escher, Bach: An Eternal Golden Braid</u>**]

The contents of phenomenal consciousness are whatever lies within the charmed spotlight of attention. The possibilities are endless: anything that we can possibly perceive, remember, or imagine is fair game. As both Edelman and Ramachandran propose, the output end is highly differentiated. But at the same time we feel as if our conscious experience is continuous and consistent with a unitary self moving through a physical universe. Order emerges out of the surrounding chaos. The coherence of the stream of consciousness is yet another illusion created by a Darwinian process. I introduced natural selection with respect to genetic evolution in chapter 6. In chapter 10, we extended Darwin's reach to the realm of memes. Now we come to a third type of natural selection: **neural Darwinism**.

Recall that a Darwinian process simply involves the replication of patterns with slight variation and the competition of these patterns for selection to the next round. This process is not limited to genes and molecules of DNA. The selection of antibodies in the immune system is another example. If we ever succeed in producing artificial intelligence or discover extraterrestrial intelligence, their evolution too will likely be governed by Darwinian type processes that have nothing to do with genes or DNA. Neuroscientists have come up with models of consciousness by applying the principles of Darwinian selection to the patterns of neural activity within those integrated cortical loops.

In a similar vein, Dan Dennett believes that the myriad contents of our conscious lives simply seem to form a continuous narrative thread because those neural patterns representing 'things' that are most cooperative or complementary to other 'things' already in the mind are biased and recruited into the spotlight of attention where they can spawn more such patterns. On the other hand, patterns that represent random things that have no existent counterparts become extinct. Chosen patterns tend to stabilize and stay in the spotlight because they attract 'like-minded' patterns. Examples include 'ham and eggs', 'give and take', 'crime and punishment', or 'unsafe sex and venereal disease'. The strength of these relationships is, in turn, determined by the usefulness of the association for the survival and reproductive success of the individual hosting these neural patterns. Once a successful 'gang' or 'consortium' of associated neural patterns representing a useful behavior or thought emerges, it tends to survive and stay in the mind. These are memes. For this reason Dennett believes that our minds and even our identities are just vast collections of cooperating memes governed by Darwinian selection. Richard Dawkins expresses this idea in <u>Unweaving the Rainbow</u>:

"The individual organism…is not fundamental to life, but something that emerges when genes, which at the beginning of evolution were separate, warring entities, gang together in cooperative groups, as 'selfish cooperators'. The individual organism is not exactly an illusion. It is too concrete for that. But it is a secondary, derived phenomenon, cobbled together as a consequence of the actions of fundamentally separate, even warring, agents…Perhaps the subjective 'I', the person that I feel myself to be, is the same kind of semi-illusion. The mind is a collection of fundamentally independent, even warring, agents. Marvin Minsky, the father of artificial intelligence, called his 1985 book <u>The Society of Mind</u>. Whether or not these agents are to be identified with memes, the point I am now making is that the subjective feeling of 'somebody in there' may be a cobbled, emergent, semi-illusion analogous to the individual body emerging in evolution from the uneasy cooperation of genes." [Dawkins, pg308-309]

Dennett maintains that the serial collections of memes illuminated within the spotlight of attention comprise 'multiple drafts' of an ongoing personal narrative. This narrative slowly emerges in early childhood and continues until death (or at least late senility) with regular breaks for (nonREM) sleep and occasionally unexpected breaks due to seizures and bumps on the head. Here are some revealing passages from his ambitious work, **Consciousness Explained** (1991):

"There is no single, definitive 'stream of consciousness', because there is no central Headquarters, no Cartesian Theater where 'it all comes together' for the perusal of a Central Meaner. Instead of such a single stream (however wide), there are multiple channels in which specialist circuits try, in parallel pandemoniums, to do their various things, creating Multiple Drafts as they go. Most of these fragmentary drafts of 'narrative' play short-lived roles in the modulation of current activity but some get promoted to further functional roles, in swift succession, by the activity of a virtual machine in the brain. The seriality of this machine (its 'von Neumannesque' character) is not a 'hard-wired' design feature, but rather the upshot of a succession of coalitions of these specialists.

"The basic specialists are part of our animal heritage. They were not developed to perform peculiarly human actions, such as reading and writing, but ducking, predator-avoiding, face-recognizing, grasping, throwing, berry-picking, and other essential tasks. They are often opportunistically enlisted in new roles, for which their native talents more or less suit them. The result is not bedlam only because the trends that are imposed on all this activity are themselves the product of design. Some of this design is innate, and is shared with other animals. But it is augmented, and sometimes even overwhelmed in importance, by microhabits of thought that are developed in the individual, partly idiosyncratic results of self-exploration and partly the predestined gifts of culture. Thousands of memes, mostly borne by language, but also by wordless 'images' and other data structures, take up residence in an individual brain, shaping its tendencies and thereby turning it into a mind." [Dennett, pg 224]

"According to our sketch, there is competition among many concurrent contentful events in the brain, and a select subset of such events 'win'. That is, they manage to spawn continuing effects of various sorts. Some, uniting with language-demons, contribute to subsequent sayings, both sayings-aloud to others and silent (and out-loud) sayings to oneself. Some lend their content to other forms of subsequent self-stimulation, such as diagramming-to-oneself. The rest die out almost immediately, leaving only faint traces—circumstantial evidence—that they ever occurred at all." [Dennett pg 275]

Dennett makes two other important points. First, not just verbal access (A consciousness) and self-awareness (S consciousness), but also the very quality of the perceptual moment—William James's 'pungency of experience'—(P consciousness) is itself an emergent property, an epiphenomenon arising from the deeper structures and patterns of neural computation and molecular biology. If and when we understand all the nuances of these deeper levels, he believes, 'qualia'—the raw feel of first person experiences—will simply evaporate much as the concept of a 'vital life force' lost support after the discovery of DNA.

Secondly, Dennett emphatically maintains that there is no central brain region where the stream of consciousness comes together. He calls this idea the 'Cartesian Theater' and accuses among others, Crick and Edelman of supporting it.

◆   ◆   ◆

I agree with Dennett's idea of multiple competing drafts of neural code producing and comprising the contents of contents of consciousness. I also accept the notion that qualia are emergent phenomena arising from more mundane brain activity (although I will preface this by saying that just because something 'emerges' doesn't mean it doesn't exist. Qualia (P consciousness) may be like that: it exists at its **own level.**

Many people have difficulty accepting the possibility that their innermost "subjective experiences either don't exist, or are unimportant because they are some sort of ambient or peripheral effect" as the maverick computer scientist Jaron Lanier puts it. **[Lanier, Sept 25, 2000,'One half of a manifesto', <u>Edge</u> <u>74</u>]** Lanier half jokingly suggests that Dan Dennett might be a freak of nature who actually lacks internal experiences, thus explaining his persistently nihilistic views on qualia.

I disagree with Dennett's final point. Attention is bound and consciousness expressed in **several key areas of the brain**. There are, in fact, many Cartesian Theaters playing nearby. In their search for the NCC, researchers have uncovered several brain areas that are likely to be very important for the binding of subjective conscious experiences. Francis Crick believes that the PFC, the thalamus, and especially the anterior cingulate gyrus are essential for consciousness. Ramachandran, on the other hand, believes that the temporal lobe is the site where awareness happens. Here is a revealing passage from Ramachandran's <u>Phantoms in the Brain</u>:

"It is surprising that many people think that the seat of consciousness is the frontal lobes, because nothing dramatic happens to qualia and consciousness per se if you damage the frontal lobes—even though the patient's personality can be profoundly altered (and he may have difficulty switching attention). I would suggest instead that most of the action is in the temporal lobes because lesions and hyperactivity in these structures are what most often produce striking disturbances in consciousness. For instance, you need the amygdala and other parts of the temporal lobes for seeing the significance of things, and surely this is a vital part of conscious experience. Without this structure you are a zombie...capable only of giving a single correct output in response to a demand, but with no ability to sense the meaning of what you are doing or saying.

"Everyone would agree that qualia and consciousness are not associated with the early stages of perceptual processing as at the level of the retina. Nor are they associated with the final stages of planning motor acts when behavior is actually out. They are associated, instead, with the intermediate stages of processing—a stage where stable perceptual representations are created (yellow, dog, monkey) and that have meaning (the infinite implications and possibilities for action from which you can choose the best one). This happens mainly in the temporal lobe and associated limbic structures, and, in this sense, the temporal lobes are the interface between perception and action.

"The evidence for this comes from neurology; brain lesions that produce the most profound disturbances in consciousness are those that generate temporal lobe seizures, whereas lesions in other parts of the brain only produce minor disturbances in consciousness. When surgeons electrically stimulate the temporal lobes of epileptics, the patients have vivid conscious experiences. Stimulating the amygdala is the surest way to 'replay' a full experience, such as an autobiographical memory or a vivid hallucination. Temporal lobe seizures are often associated not only with alterations in consciousness in the sense of personal identity, personal destiny and personality, but also with vivid qualia—hallucinations such as smells and sounds. If these are mere memories, as some claim, why would the person say, 'I literally feel like I'm reliving it'? These seizures are characterized by the vividness of the qualia they produce. The smells, pains, tastes and emotional feelings—all generated in the temporal lobes—suggest that this brain region is intimately involved in qualia and conscious awareness.

"Another reason for choosing the temporal lobes—especially the left one—is that this is where much of language is represented. If I see an apple, temporal lobe activity allows me to apprehend all its implications almost simultaneously. Recognition of it as a fruit of a certain type occurs in the inferotemporal cortex, the amygdala gauges the apple's significance for my well-being and Wernicke's and other areas alert me to all the nuances of meaning that the mental image—including the word 'apple'—evokes; I can eat the apple, I can smell it; I can bake a pie, remove its pith, plant its seeds; use it to 'keep the doctor away', tempt Eve and on and on. If one enumerates all of the attributes that we usually associate with the words 'consciousness' and 'awareness', each of them, you will notice, has a correlate in temporal lobe seizures, including vivid visual and auditory hallucinations, 'out of body' experiences and an absolute sense of omnipotence or omniscience. Any one of this long list of disturbances in conscious experience can occur individually when other parts of the brain are damaged (for instance, disturbances of body image and attention in parietal lobe syndrome), but it's only when the temporal lobes are involved that they occur simultaneously or in different combinations; that again suggests that these structures play a central role in human consciousness." [Ramachandran pg 244-246]

◆   ◆   ◆

The picture that is emerging is that the three different species of consciousness are distinct and potentially dissociable. **P consciousness**, with its irrevocable qualia, often colored by an emotional tag, comes from the temporal lobe (including the ventral visual stream) and the limbic structures of the hippocampus and amygdala. **S consciousness**, with its self reflective 'I' quality, requires much of the machinery of P consciousness **plus** the parietal cortex, where several body-centered and object-centered maps are stored. **A consciousness**, which depends on language, utilizes the left hemisphere language areas. All of these activated cortical areas are in close contact with the thalamus (the 'loops') to produce the roving spotlight of attention.

Inside each of us, there are multiple little Cartesian Theaters, each with its own separate modules and special access to the thalamus. All of these specialized thalamocortical circuits are continuously playing multiple drafts of one's ongoing conscious life. Just like a major sports broadcast where several cameras in various locations are focused on different aspects of the subject and the producers actively shift from one view to another, so it is with consciousness. Attention, biased both from the top down and from bottom up, is the 'producer' that chooses which theater will broadcast conscious experience. If attention's camera shines from the temporal lobe, we become aware of the raw, elemental quality of objects and situations. If it comes from the parietal lobe, we become aware of our embodied physical self. If it comes from the left frontal lobe, we are able to articulate feelings into words and speech. Through it all, there emerges a sense of autobiographical unity: a 'metaconjunction', if you will, not of features in a visual setting, but of the composite aspects that make up one's identity. Antonio Damasio writes about this elegantly in his landmark book, **<u>Descartes' Error</u>**:

"…In using the notion of self, I am in no way suggesting that **all** the contents of our minds are inspected by a single central knower and owner, and even less that such an entity would reside in a single brain place. I am saying, though, that our experiences tend to have a consistent perspective, as if there were indeed an owner and knower for most, though not all, contents. I imagine this perspective to be rooted in a relatively stable, endlessly repeated biological state. The source of the stability is the predominantly invariant structure and operation of the organism, and the slowly evolving elements of autobiographical data." (pg. 238)

"At each moment the state of self is constructed, from the ground up. It is an evanescent reference state, so continuously and consistently **re**constructed that the owner never knows it is being **re**made unless something goes wrong with the remaking. The background feeling now, or the feeling of an emotion now, along with the non-body sensory signals now, happen to the concept of self as instantiated in the coordinate activity of multiple brain regions. But our self, or better even, our metaself, only 'learns' about that 'now' an instant later…Present continuously becomes past, and by the time we take stock of it we are in another present, consumed with planning the future, which we do on the stepping-stones of the past. The present is never here. We are hopelessly late for consciousness." (pg. 240)

Free Will

At this point, we should briefly touch on the concept of free will. Scholars have endlessly debated its very definition and I don't want to get too embroiled in the ancient controversy, but I do have some personal thoughts on the matter. First of all, the 'freedom of choice' invoked by free will, whether to tap our toes or pull the trigger of a gun, involves analysis a level apart from physics: the old puzzle of how freedom is compatible with a deterministic or a chaotic universe. The cosmos, as governed by physical laws (thermodynamics, statistics, quantum mechanics), is either determined or not to some degree. On the one hand, it is at least conceivable that knowledge of all particles and forces at a given time zero allows one to predict with some precision the state of the universe at a subsequent time zero plus one. In actuality, this is impossible (even in theory) because of the quantum uncertainty 'built' into the very fabric of space-time. Thus there is a degree of randomness and unpredictability in the system. The universe id both determined (at large scales), and random (at small scales of measurement). David Hume has famously pointed out that neither situation is compatible with free will. But all this debate is at the level of physics, or what Dan Dennett calls the 'physical stance'.

The freedom to choose a particular action or thought is a function of brains, and brains are complex machines **designed** by Natural Selection. Although all machines, be they brains, bodies, or computers, are subject to basic physical laws (statistics and quantum mechanics), at a fundamental level, they are **better explained** by the laws of design (neuroscience, biology, computer logic). This is all quite obvious to the biologist who would never attempt to use quantum physics to understand the behavior of a cell, or to the political scientist who would never imagine using cell biology to explain the behavior of social groups.

At the level of design, a type of freedom does emerge. It is the multiplicity of possibility. A machine can behave in unpredictable ways. The more complex the design, the more **interesting** the possible outcome. There are only a finite number of interesting ways a thrown stone can fall. But a bacterial cell has rather more freedom of action. And a human being has even more still. The point is not what the final outcome actually turns out to be (there will always be one final outcome—in that sense, we do live in a determined universe), but rather how many different interesting outcomes are possible given the starting conditions. Physics determines the final outcome, but design from selection determines the scope of possibility.

So freedom comes from design. But this freedom is not the same as 'free will'. Will means more than just possibility, it implies **choice.** The choice may or may not be conscious. In fact, we make unconscious choices all the time. Motor automatisms such as blinking and lip smacking are often unconscious choices. More to the point, the vast majority of the activity in your brain at this very moment is beyond the reach of conscious awareness. These are the myriad parallel streams of neural pattern coursing through the circuits of the basal ganglia, thalamus, hypothalamus, amygdala, and so on that make up the model mind. As subconscious as all this activity is, it controls much (and likely most) of our behavior. This is not surprising given that consciousness is likely a recent development in animal evolution. Most other species of life on earth do extremely well with their (subconscious) choices, thank you.

It is an illusion that we are actually in control of most of our actions. Benjamin Libet has done some very intriguing experiments measuring brain activity in subjects just prior to 'voluntary' behavior. The conscious intention to commit a voluntary act (flicking the wrist) occurs about 200 milliseconds prior to muscle action. But as much as 800 milliseconds prior to this (a full second prior to the act) there is measurable brain activity in the premotor and frontal cortex corresponding to the action. This has been called the 'readiness potential' or RP. Libet believes that the RP is the subconscious signature of all voluntary activity. If so, then we are never really in conscious control of our actions. There is no such thing as free will. But Libet went on to discover that RPs also preceded subsequently aborted actions. In other words, we can **consciously veto** subconsciously chosen actions. This has prompted Ramachandran to joke that although free will is an illusion, 'free won't' may be real!

If there really is a limited type of conscious free will, the kind that at least has veto power over subconscious decisions and actions, then I think it belongs in the spotlight of attention. In fact, the very contents of the 2-D circular spotlight extended into the 3-D cylinder of my mathematical stream of consciousness is a **reflection** and **interpretation** of the subconscious neural activity swirling around it. The size of the spotlight, the area under the focus of attention, is a function of activity in the PFC and the anterior cingulate. Continuing the camera analogy, the aperture size fluctuates with time and attention, corresponding to the contents of one's stream of consciousness. As Marcel Kinsbourne put it, '...at any time in the awake individual there is a **dominant focus** of coordinated patterned neural activity that underlies the phenomenal experience of that moment, the momentarily dominant draft of the multiple drafts that in rapid succession constitute the apparently continuous stream of ever-changing consciousness.' This stream simultaneously tracks the subconscious neural activity around it, but at a distance.

To summarize, freedom of a kind does exist regardless of whatever degree of determinism or randomness rules the universe. This is because freedom of choice is predicated on the range of interesting possibilities allowed by design from selection. But the choices and actions brought about by this freedom is largely unconsciousness: the end result of unmonitored brain activity. Consciousness, on the other hand, is a separate and, in some cases, separable process **mapped onto** all this underlying activity. Conscious will does not cause action; rather it continuously monitors and interprets the ongoing subconscious processes that cause what we do. And importantly, it has veto power to restrain otherwise automatic behavior. The psychologist Daniel Wegner has proposed a neat synthesis of these ideas:

The experience of will is the way our minds portray their operations to us, then, not their actual operation...The real causal mechanism is the marvelously intricate web of causation that is the topic of scientific psychology...the real causal mechanisms underlying behavior are never present in consciousness. Rather, the engines of causation are unconscious mechanisms of mind...The unique human convenience of conscious thoughts that preview our actions gives us the privilege of feeling we willfully cause what we do. In fact, unconscious and inscrutable mechanisms create both conscious thought about action and create the action as well, and also produce the sense of will we experience by perceiving the thought as the cause of action. So, although our thoughts may have deep, important, and unconscious causal connections to our actions, the experience of conscious will arises from a process that interprets these connections, not from the connections themselves...**[Wegner, D.M. & Wheatley, T. (1999)'Apparent Mental Causation: sources of the experience of will'** <u>American Psychologist</u>, 54, 480-492]

Natural Selection has seen to it that the stream of consciousness closely tracks the subconscious/subcortical streams of neural activity that actually cause our behavior. The tightness of this fit is the reason we usually feel so in control of ourselves. However, when the fit slips from fatigue or mental illness (such as schizophrenia or autism), one may no longer feel that one is in command of one's thoughts, intentions, and actions.

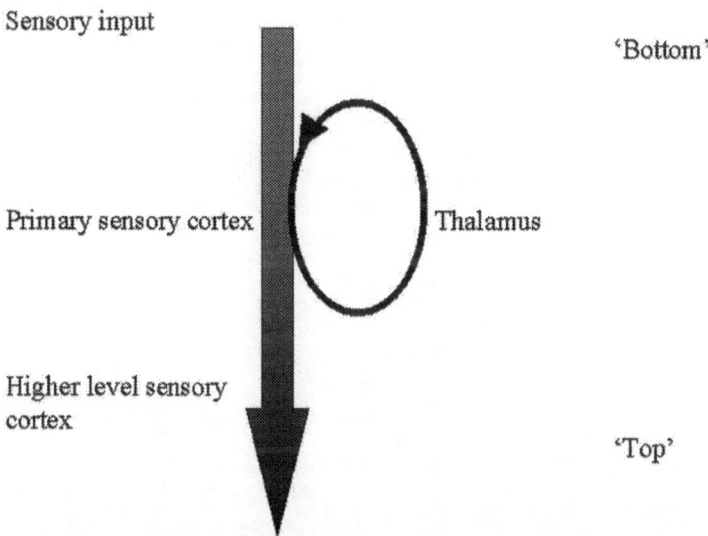

Figure 1: Perception/thalamo-cortical circuits

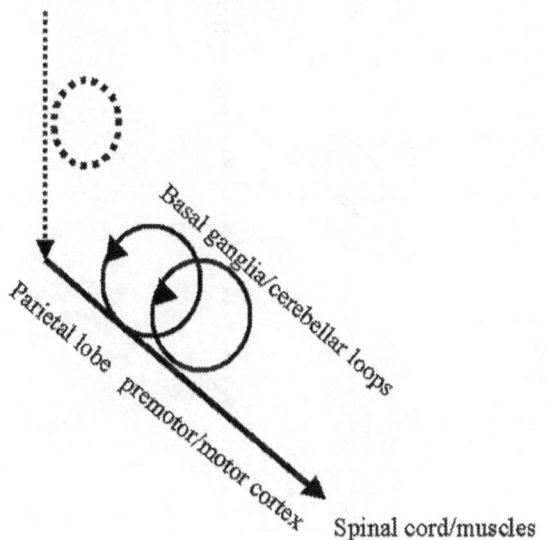

Figure 2: Motion

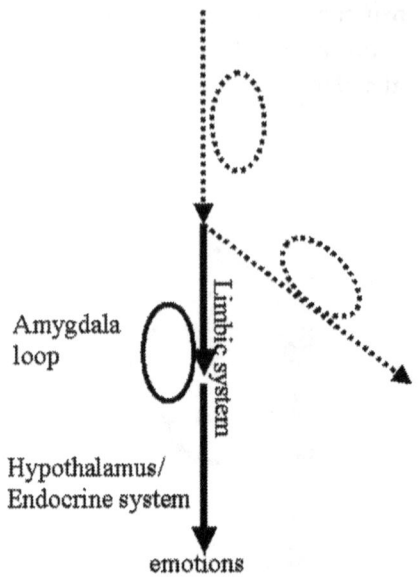

Figure 3: Emotion

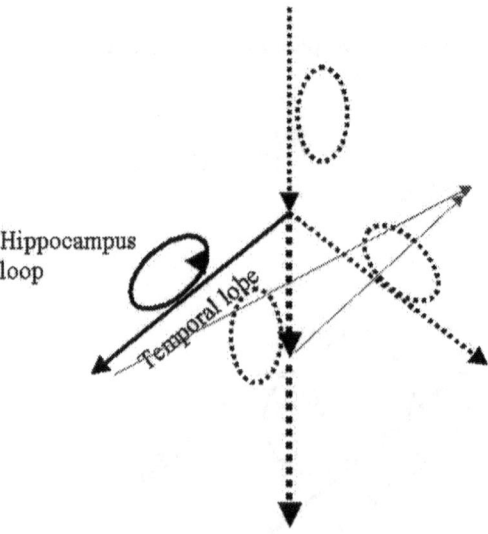

Figure 4: Memory

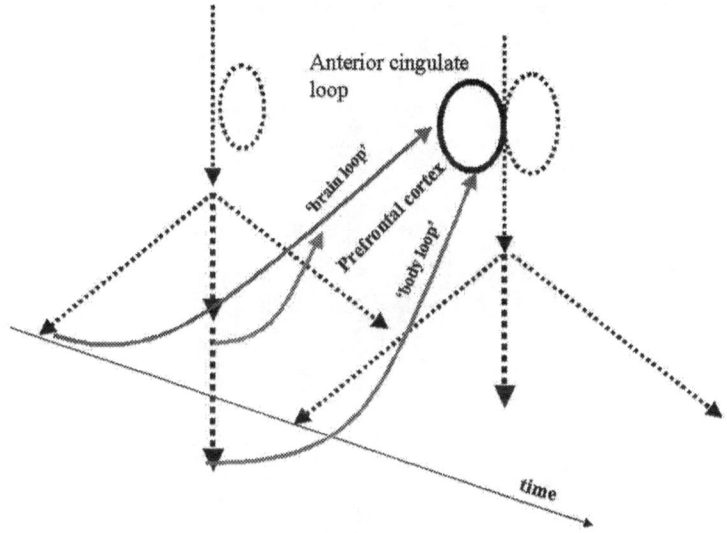

Figure 5: Executive control

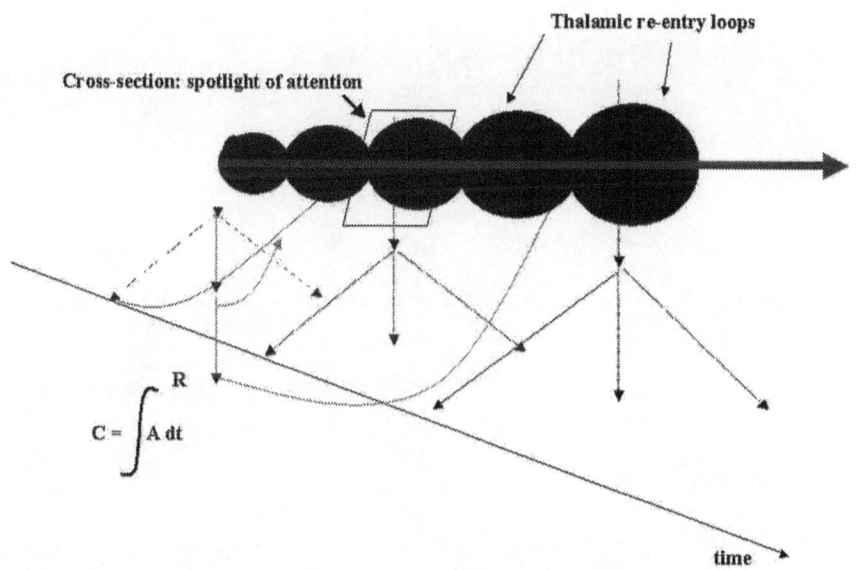

Figure 6: A Model Mind

978-0-595-39296-4
0-595-39296-2

www.ingramcontent.com/pod-product-compliance
Lightning Source LLC
Chambersburg PA
CBHW030254290526
45785CB00001B/87